Essential
AS Physics
for OCR

Jim Breithaupt

Published in 2004 by:
Nelson Thornes Ltd
Delta Place
27 Bath Road
CHELTENHAM
GL53 7TH
United Kingdom

04 05 06 07 08 / 10 9 8 7 6 5 4 3 2 1

A catalogue record for this book is available from the British Library

ISBN 0 7487 8507 8

Illustrations by IFA Design
Page make-up by Tech-Set Ltd

Printed and bound in Spain by Graficas Estella

Contents

Module 3 (Part 1) Wave properties

Module 3 (Parts 2–3) Coursework and practical examination

Introduction

This book is written for AS Physics' students who are following or about to follow OCR Physics (Specification A). It provides the entire content of all three AS modules as well as information and advice about practical assessment through both the practical examination and coursework. In addition, chapters on essential maths skills and study skills are provided. Guidance on how to bridge the gap between Physics at GCSE and at AS level is given below.

About AS Physics

Welcome to Physics at AS level, the subject that helps us to understand nature from the smallest possible scale (deep inside the atom) to the largest conceivable scale (stretching across the entire Universe). Physics is about making predictions, testing them through observations and measurements, and devising theories and laws to make more predictions. On your course, you will cover the key ideas of the subject and you will learn about topics such as mechanics, electricity and waves. You will learn the skills of making observations and measurements, and how to use your mathematical skills to make sense of experiments. You will also learn how to communicate effectively your knowledge and understanding of the subject. You will discover that Physics is a very creative subject that calls for imagination and inventiveness. Don't be afraid to ask your teacher when something doesn't seem clear. Einstein developed a reputation for asking awkward questions and it made him into the most famous physicist ever! Even a century after he discovered the famous equation $E = mc^2$, we still don't have a clear understanding of how this equation works.

So we are now about to set off on a course that will further develop your knowledge, understanding and skills of the subject. At this stage, your GCSE course will have provided you with solid foundations to build on as you progress through the course. To help you make a smooth transition from GCSE to AS physics, Skills for starting AS Physics, p.6, covers the key mathematical and practical skills you need to get started on your course. Further support on mathematical and practical skills is given in Chapters 12 and 13. Chapter 13 also includes advice on coursework as well as help in preparing for the practical examination. In Chapter 14, you will find useful advice to help you as you progress beyond the initial stages of your course and as you revise and prepare for exams.

This book is written for the AS Physics course for OCR Specification A; and matches the specification. The main chapters of the book, Chapters 1–11, are written in sequence to follow each of the three Modules (or units) that comprise the AS course. Short questions are supplied at the end of each topic and longer examination-style questions at the end of each chapter. The three Modules are listed below, together with the chapters in this book that are written for each module. At the end of the book, there is a reference section consisting of useful data and formulae, a glossary, answers to numerical questions and a comprehensive index. A full list of the formulae you need to remember for your exams is given, as well as a further list of the formulae that you will be provided with in each examination paper. Essential practical experiments are provided in the book as appropriate. Further practical experiments are provided as part of the Teacher Support Pack.

AS Course structure

Module 1	**Forces and motion**	Chapters 1–5
Module 2	**Electrons and photons**	Chapters 6–9
Module 3	**Part 1 Wave properties**	Chapters 10–11
	Parts 2–3 (Coursework and practical examination)	Chapter 13

This book is written to help you pass the AS examination – so make sure you use it fully, particularly the final 'skills' chapters and the final reference section. It is also written to provide you with a firm foundation for your A2 course using the companion book *Essential A2 Physics* written for OCR Specification A. Using this AS book will help you to pass your AS exam and then to move on successfully to A2 Physics. More importantly, I hope it will give you a lasting interest in Physics and an on-going enthusiasm for this exciting subject.

Jim Breithaupt

Skills for starting AS Physics

1. Using a calculator

Practice makes perfect when it comes to using a calculator. For AS Physics, you need no more than a scientific calculator. You need to make sure you master the technicalities of using a scientific calculator as early as possible in your AS Physics course. At this stage, you should be able to use a calculator to add, subtract, multiply, divide, find squares and square roots and calculate sines, cosines and tangents of angles. Further important calculator functions are described in section 12.1.

2. Making measurements

You should know at this stage how to make measurements using basic equipment such as metre rules, protractors, stopwatches, thermometers, balances (for weighing an object) and ammeters and voltmeters. During the course, you will also be expected to use micrometers, verniers, oscilloscopes and data loggers. The use of these items is described for reference in section 13.4.

Here are some useful reminders about making measurements:

- Check the **zero reading** when you use an instrument to make a measurement. For example, a metre ruler worn away at one end might give an error.
- When a multi-range instrument is used, start with the **highest range** and switch to a lower range if the reading is too small to measure accurately.
- Make sure you record all your measurements in a **logical order**, stating the correct unit of each measured quantity.
- Don't pack equipment away until you are sure you have enough measurements or you have checked unexpected readings. See section 13.5 for **anomalous** measurements.

3. Using measurements in calculations

Whenever you make a record of a measurement, you must always note the correct unit as well as the numerical value of the measurements.

The scientific system of units is called the SI system. This is described in more detail in section 12.1. The base units of the SI system you need to remember are listed below. All other units are derived from the SI base units.

- the metre (m) is the SI unit of length. Note also that 1 m = 100 cm = 1000 mm.
- the kilogram (kg) is the SI unit of mass. Note that 1 kg = 1000 g.
- the second (s) is the SI unit of time.
- the ampere (A) is the SI unit of current.

Powers of ten and numerical prefixes are used to avoid unwieldy numerical values. For example,

- $1\,000\,000 = 10^6$ which is 10 raised to the power 6 (usually stated as '10 to the 6').
- $0.000\,000\,1 = 10^{-7}$ which is 10 raised to the power -7 (usually stated as '10 to the -7').

Prefixes are used with units as abbreviations for powers of ten. For example, a distance of 1 kilometre may be written as 1000 m or 10^3 m or 1 km. The most common prefixes are shown in Table A.

Table A Prefixes

Prefix	pico-	nano-	micro-	milli-	kilo-	mega-	giga-	tera-
Value	10^{-12}	10^{-9}	10^{-6}	10^{-3}	10^3	10^6	10^9	10^{12}
Prefix symbol	p	n	μ	m	k	M	G	T

Note that the cubic centimetre (cm³) and the gram (g) are in common use and are therefore allowed as exceptions to the prefixes in Table A.

Standard form is usually used for numerical values smaller than 0.001 or larger than 1000.

- The numerical value is written as a number between 1 and 10 multiplied by the appropriate power of ten. For example:

$$64\,000\,\text{m} = 6.4 \times 10^4\,\text{m}$$
$$0.000\,005\,1\,\text{s} = 5.1 \times 10^{-6}\,\text{s}$$

- A prefix may be used instead of some or all of the powers of ten. For example,

$$35\,000\,\text{m} = 35 \times 10^3\,\text{m} = 35\,\text{km}$$
$$0.000\,000\,59\,\text{m} = 5.9 \times 10^{-7}\,\text{m} = 590\,\text{nm}$$

To convert a number to standard form, count how many places the decimal point must be moved to make the number between 1 and 10. The number of places moved is the power of ten that must accompany the number between 1 and 10. Moving the decimal place:

- to the *left* gives a *positive* power of ten (e.g. $64\,000 = 6.4 \times 10^4$).
- to the *right* gives a *negative* power of ten (e.g. $0.000\,005\,1 = 5.1 \times 10^{-6}$).

4. Using the right-angled triangle

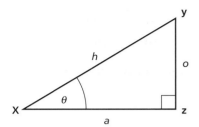

Fig B *A right-angled triangle*

The sine, cosine and tangent of an angle are defined from the right-angled triangle. Figure B shows a right-angled triangle XYZ, in which side XY is the **hypotenuse** (i.e. the side opposite the right angle), side YZ is **opposite** angle θ and side XZ is **adjacent** to angle θ.

$$\sin \theta = \frac{YZ}{XY} = \frac{o}{h} \quad \text{where } o = YZ, \text{ the side opposite angle } \theta$$

$$\cos \theta = \frac{XZ}{XY} = \frac{a}{h} \qquad h = XY, \text{ the hypotenuse}$$

$$\tan \theta = \frac{YZ}{XZ} = \frac{o}{a} \qquad a = XZ, \text{ the side adjacent to angle } \theta$$

Pythagoras' theorem states that for any right-angled triangle:

**The square of the hypotenuse
= the sum of the squares of the other two sides**

Applying Pythagoras' theorem to the right-angled triangle XYZ gives:

$$\mathbf{XY^2 = XZ^2 + YZ^2}$$

5. Using equations

Symbols are used in equations and formulae to represent physical variables. In your GCSE course, you may have used equations with words instead of symbols to represent physical variables. For example , you will have met the equation 'distance moved = speed × time'. Perhaps you were introduced to the same equation in the symbolic form

'$s = vt$', where s is the symbol used to represent distance , v is the symbol used to represent speed and t is the symbol used to represent time. The equation in symbolic form is easier to use because the rules of algebra are more easily applied to it than to a word equation. In addition, writing words in equations wastes valuable time. However, you need to remember the agreed symbol for each physical quantity.

Equations often need to be rearranged. This can be confusing if you don't learn the following basic rules at an early stage:

- **Read an equation properly.** For example, the equation $v = 3t + 2$ is not the same as the equation $v = 3(t + 2)$. If you forget the brackets when you use the second equation to calculate v when $t = 1$, then you will get $v = 5$ instead of the correct answer $v = 9$. The first equation tells you to multiply t by 3 then add 2. The second equation tells you to add t and 2, then multiply the sum by 3.
- **Rearrange an equation properly.** In simple terms, always make the same change to both sides of an equation. For example, to make t the subject of the equation $v = 3t + 2$,

Step 1: Subtract 2 from both sides of the equation,

$$\text{so} \qquad v - 2 = 3t + 2 - 2$$
$$= 3t$$

Step 2: The equation is now $v - 2 = 3t$ and can be written $3t = v - 2$

Step 3: Divide both sides of the equation by 3,

$$\text{so} \qquad \frac{3t}{3} = \frac{v - 2}{3}$$

Step 4: Cancel 3 on the top and the bottom of the left-hand side, to finish with $t = \dfrac{v - 2}{3}$

To use an equation as part of a calculation:

- Start by making the quantity to be calculated the **subject** of the equation.
- Write the equation out with the **numerical values** in place of the symbols.
- Carry out the calculation and make sure you give the answer with the **correct unit**.

Unless the equation is simple (e.g. $V = IR$), don't insert numerical values then rearrange the equation. Rearrange, *then* insert the numerical values; you are less likely to make an error if the numbers are inserted later in the process.

MODULE ONE

1

Forces and motion

On the move

1.1

Vectors and scalars

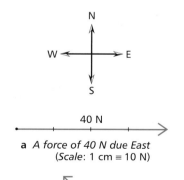

Fig 1.1 *Map of locality*

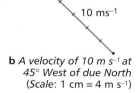

a *A force of 40 N due East*
(*Scale*: 1 cm ≡ 10 N)

b *A velocity of 10 m s⁻¹ at 45° West of due North*
(*Scale*: 1 cm ≡ 4 m s⁻¹)

Fig 1.2 *Representing a vector*

Imagine you are planning to cycle to a friend's home several kilometres away from your home. The **distance** you travel depends on your route. However, the direct distance from your home to your friend's home is the same whichever route you choose. Distance in a given direction or **displacement** is an example of a **vector** quantity because it has magnitude and direction.

> **A vector is any physical quantity that has a direction as well as a magnitude.**

Further examples of vectors include velocity, acceleration, force and weight.

> **A scalar is any physical quantity that is not directional.**

For example, distance is a scalar because it takes no account of direction. Further examples of scalars include mass, density, volume and energy.

1.1.1 Representing vectors

Any vector can be represented as an arrow. The length of the arrow represents the **magnitude** of the vector quantity. The direction of the arrow gives the **direction** of the vector.

- **Displacement** is distance in a given direction. The displacement from one point to another can be represented on a map or a scale diagram as an arrow from the first point to the second point. The length of the arrow must be in proportion to the least distance between the two points.
- **Velocity** is speed in a given direction. The velocity of an object can be represented by an arrow of length in proportion to the speed pointing in the direction of motion of the object.
- **Force and acceleration** are both vector quantities and therefore can each be represented by an arrow in the appropriate direction and of length in proportion to the magnitude of the quantity.

On a journey

Cyclists and hill walkers should always take a map and compass to make sure they do not get lost. A compass tells the user which direction is North. A map tells the user how far he or she has gone. Consider the map shown in Figure 1.1. Suppose your home is at O and your friend's home is at A. Your route-plan is to cycle along the road heading North, over the railway bridge, then turn East at the next road junction.

- The **distance** to be cycled can be estimated by measuring the length of the route on the map in centimetres, then using the map scale.
- The **displacement** or direct distance from O to A is marked on the map as an arrow OA.

- Your **velocity** at any point on your journey changes, because you change direction and because your speed changes. Suppose your speed as you pass over the railway bridge is 2.0 m s^{-1}. The direction in which you are travelling as you pass over the bridge is about 10° East of due North. You can check this on Figure 1.1 using a protractor. So your velocity at the railway bridge is 2.0 m s^{-1} in a direction which is 10° East of due North.

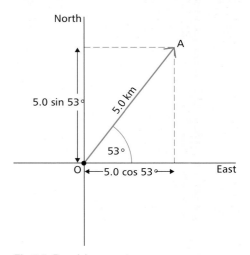

1.1.2 Vector components

The displacement vector **OA**, shown on Figure 1.1, can be represented:

- as an arrow of length in proportion to the direct distance of 5.0 km from O to A. The direction of the arrow would be 53° North of due East.
- as a map reference, one part stating how far A is East or West of O and the other stating how far A is North or South of O. The two parts of the map reference, referred to as **components**, may be written as (3.0 km, 4.0 km) where East/West is first. This is the same as writing the **coordinates** of a point on a graph as (x, y), where x is the distance from the origin along the x-axis and y is the distance from the origin along the y-axis.

Fig 1.3 *Resolving a vector*

Resolving a vector into two perpendicular components

Resolving a vector is the process of working out the components of a vector in two perpendicular directions from the magnitude and direction of the vector. Figure 1.3 shows the displacement vector OA represented on a scale diagram that also shows lines due North and due East. The components of this vector along these two lines are 5.0 cos 53° km (= 3.0 km) along the line due East and 5.0 sin 53° km (= 4.0 km) along the line due North.

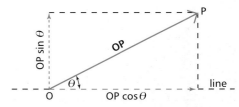

Fig 1.4 *The general rule for resolving a vector*

In general, to resolve any vector into two perpendicular components, draw a diagram showing the two perpendicular directions and an arrow to represent the vector. Figure 1.4 shows this diagram for a vector OP. The components are represented by the projection of the vector onto each line. If the angle θ between the vector OP and one of the lines is known:

- the component along that line = OP cos θ, and
- the component perpendicular to that line (i.e. along the other line) = OP sin θ.

Example

An aircraft in level flight has a constant velocity of 50 m s^{-1} in a direction of 40° North of due East, as shown in Figure 1.5. The angle between the direction of its velocity and due East is therefore 40°.

Its components of its velocity are, in m s^{-1},

- 50 cos 40° due East, and
- 50 sin 40° due North.

1.1.3 Addition of vectors

Using a scale diagram

Let's go back to the cycle journey in Figure 1.1. Suppose when you reach your friend's home at A, you then go on to another friend's home at B. Your journey is now a two-stage journey:

- **Stage 1, from O to A,** is represented by the displacement vector **OA**.
- **Stage 2, from A to B,** is represented by the displacement vector **AB**.

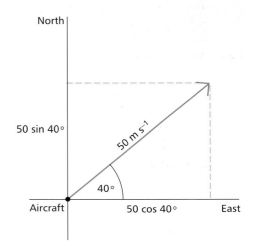

Fig 1.5 *Resolving velocity*

Figure 1.6 shows how the overall displacement from O to B, represented by vector **OB**, is the result of adding vector **AB** to vector **OA**. The **resultant** is the third side of a triangle, where OA and AB are the other two sides.

$$\mathbf{OB} = \mathbf{OA} + \mathbf{AB}$$

Use Figure 1.1 to show that the resultant displacement **OB** is 5.1 km in a direction 11° North of due East.

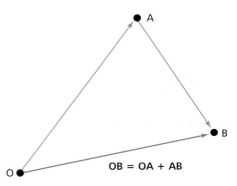

Fig 1.6 *Displacement from O to B*

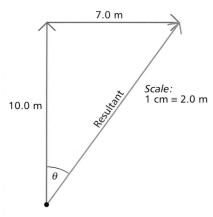

Fig 1.7 *Adding two displacements at right angles to each other*

Examples

1 Adding two displacement vectors that are at right angles to each other

Suppose you walk 10.0 m forwards, then turn through exactly 90° and walk 7.0 m. At the end, how far will you be from your starting point? The vector diagram to add the two displacements is shown in Figure 1.7, drawn to a scale of 1 cm to 2.0 m. The two displacements form two sides of a right-angled triangle, with the resultant as the hypotenuse. Using a ruler and a protractor, the resultant displacement can be shown to be a distance of 12.2 m at an angle of 35° to the initial direction. You can check this using:

- Pythagoras' theorem for the distance ($= (10.0^2 + 7.0^2)^{1/2}$), and
- the trigonometry equation $\tan \theta = \dfrac{7.0}{10.0}$ for the angle θ between the resultant and the initial direction.

2 Two forces acting at right angles to each other

Figure 1.8 shows an object, O, acted on by two forces, 7.0 N and 10.0 N, at right angles to each other. The vector diagram for this situation is also shown.

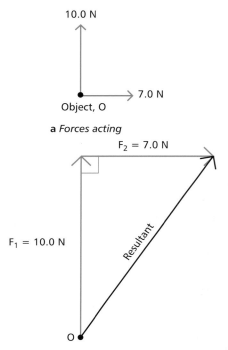

Fig 1.8 *Two forces acting at right angles to each other*

a *Forces acting*

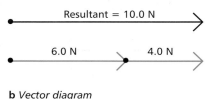

b *Vector diagram*

Fig 1.9 *Two forces acting in the same direction*

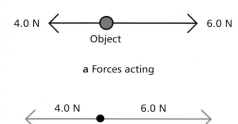

b *Vector diagram*

Fig 1.10 *Two forces acting in opposite directions*

The resultant of a force of 10.0 N acting at right angles to a force of 7.0 N has a magnitude of 12.2 N in a direction of 35° to the direction of the force of 10.0 N. You could draw a scale diagram to check this, or use Pythagoras' theorem and the appropriate trigonometry equation. You might, however, recognise that you have already done this! See Example 1, if necessary.

Note The resultant of two vectors that act along the same line has a magnitude that is:

- The **sum,** if the two vectors are in the *same* direction. For example, if an object is acted on by a force of 6.0 N and a force of 4.0 N, both acting in the same direction, the resultant force is 10.0 N.
- The **difference,** if the two vectors are in *opposite* directions. For example, if an object is acted on by a 6.0 N force and a 4.0 N force in opposite directions, the resultant force is 2.0 N in the direction of the 6.0 N force.

QUESTIONS

1 A ship leaves a port, A, and travels a distance of 60 km due North; it then changes direction at B and travels a distance of 20 km due East to a second port, C. Calculate:

a the direct distance from A to C,

b the angle, θ, between due North and the line from A to C.

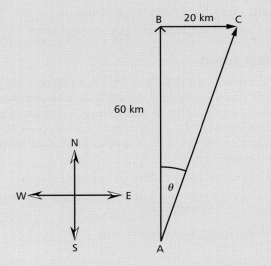

Fig. 1.11

2 An aircraft travels on a straight flight path at a constant speed of 80 m s^{-1}, in a direction 60° North of due East:

a Calculate the components of the aircraft's velocity due North and due East.

b Calculate how far North the aircraft travels in 300 seconds.

3 A crane is used to raise one end of a steel girder off the ground, as shown in Figure 1.12. When the cable attached to the end of the girder is at 20° to the vertical, the force of the cable on the girder is 6.5 kN. Calculate the horizontal and vertical components of this force.

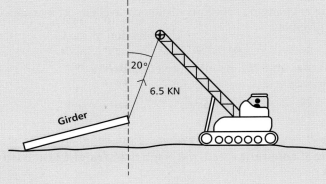

Fig. 1.12

4 On an orienteering exercise, a team is told to go from a car park, C, to a meeting point, P, 12 km away in a direction which is 30° North of due East, as shown in Figure 1.13.

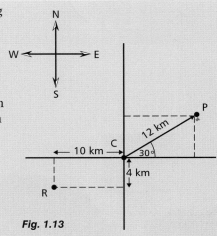

Fig. 1.13

a Calculate how far P is: (i) due East of C, (ii) due North of C.

b A railway station, R, is 10 km due West and 4 km due South of car park C.

Calculate how far P is: (i) due East of R, (ii) due North of R.

c Calculate the direct distance from P to R.

5 Figure 1.14 shows three situations, **a–c**, where an object is acted on by two known forces. For each situation, calculate the magnitude and direction of the resultant force.

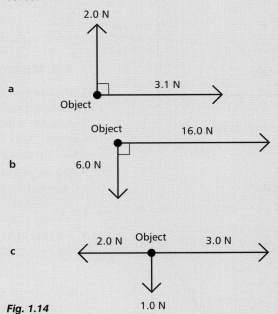

Fig. 1.14

6 Calculate the magnitude and direction of the resultant force on an object that is acted on by a force of 4.0 N and a force of 10 N that are:

a in the same direction

b in opposite directions

c at right angles to each other.

Speed and velocity

Fig 1.15

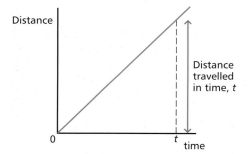

Fig 1.16 *Constant speed*

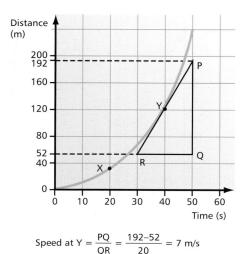

Speed at Y = $\dfrac{PQ}{QR}$ = $\dfrac{192-52}{20}$ = 7 m/s

Fig. 1.17 *Changing speed*

Fact file

- **Speed** is defined as change of distance per unit time.
- **Velocity** is defined as change of displacement per unit time. In other words, velocity is speed in a given direction.
- Speed is a **scalar** quantity. Velocity is a **vector** quantity.
- The unit of speed and of velocity is the **metre per second (m s⁻¹)**.

1.2.1 Motion at constant speed

On a motorway journey, the distance a vehicle travels in a certain time can easily be worked out if the vehicle speed is constant during that time. An object moving at a constant speed travels equal distances in equal times. For example, a car travelling at a speed of 30 m s⁻¹ on a motorway travels a distance of 30 m every second or 1800 m every minute. In 1 hour, the car would therefore travel a distance of 108 000 m or 108 km.

- For an object which travels distance *s*, in time *t* at constant speed,

$$\textbf{Speed, } v = \frac{s}{t}$$

$$\textbf{Distance travelled, } s = vt$$

- For an object moving at constant speed on a circle of radius *r*, its speed

$$v = \frac{2\pi r}{T}$$

where *T* is the time to move round once and $2\pi r$ is the circumference of the circle.

1.2.2 Motion at changing speed

There are two types of speed cameras. One type measures the speed of a vehicle as it passes the camera. The other type is linked to a second speed camera and a computer, which works out the average speed of the vehicle between the two cameras. This will catch drivers who slow down for a speed camera and then speed up again!

- For an object that travels a distance *s* in time *t*,

$$\textbf{Average speed, } v_{av} = \frac{s}{t}$$

- For an object moving at changing speed, its distance travelled, Δs, in a short time interval, Δt, is given by:

$$\Delta s = v \, \Delta t$$

where *v* is the speed at that time (i.e. its instantaneous speed). Rearranging this equation gives:

$$v = \frac{\Delta s}{\Delta t}$$

Distance-time graphs

- For an object moving at **constant** speed, the graph is **a straight line with a constant gradient**.

$$\textbf{Speed of the object} = \frac{\textbf{distance travelled}}{\textbf{time taken}} = \textbf{gradient of the line}$$

- For an object moving at **changing** speed, **the gradient of the line changes**. The gradient of the line at any point can be found by drawing a tangent to the line at that point and then measuring the gradient of the tangent. This is shown in Figure 1.17, where PR is the tangent at point Y on the line. Show for yourself that the speed at point X on the line is 2.7 m s^{-1}.

1.2.3 Velocity

Take care when cycling or driving along a country lane. Such roads can be deceptive, because they wind round 'blind bends' where drivers are unable to see oncoming traffic. In such circumstances, a driver has to be ready to change speed, and change direction, very rapidly. The velocity of a car on a country road is very unlikely to be constant!

- An object moving at constant velocity moves at the same speed without changing its direction of motion. In other words, an object moving at constant velocity travels along a straight line, covering equal distances in equal times.
- If an object changes or reverses its direction of motion, its velocity changes. For example, the velocity of an object moving on a circular path at constant speed changes continuously, because its direction of motion changes continuously (Fig 1.18).
- An object travelling along a straight line has two possible directions of motion. To distinguish between the two directions, we need a direction code where + values are in one direction and − values in the opposite direction.

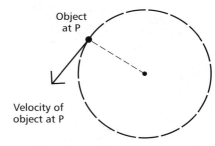

Fig 1.18 Circular motion

QUESTIONS

1 kilometre (km) = 1000 metres(m)

1 A car travels a distance of 60 km in 45 minutes at constant speed. Calculate its speed in:
   a km h^{-1}, **b** m s^{-1}.

2 A satellite moves round the Earth at constant speed on a circular orbit of radius 8000 km, with a time period of 120 minutes. Calculate its orbital speed in:
   a km h^{-1}, **b** m s^{-1}.

3 A vehicle joins a motorway and travels at a steady speed of 25 m s^{-1} for 30 minutes, then it travels a further distance of 40 km in 20 minutes before leaving the motorway (Fig 1.19). Calculate:
   a the distance travelled in the first 30 minutes,
   b its average speed on the motorway.

4 **a** Explain the difference between speed and velocity.

   b A police car joins a straight motorway at Junction 4 and travels for 12 km at a constant speed for 400 s; then it leaves at Junction 5 and rejoins on the opposite side, and travels for 8 km at a constant speed for 320 s to reach an accident(Fig 1.20). Calculate:
   (i) the displacement from Junction 4 to the accident,
   (ii) the velocity of the car on each side of the motorway.

   c Sketch a displacement-time graph for the journey.

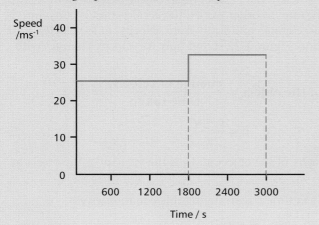

Fig 1.19

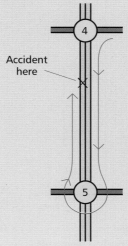

Fig 1.20

Acceleration

Fig 1.21

The acceleration of the new model above is 0.3 m s^{-2}, because its speed increased by 0.3 m s^{-1} every second.

The acceleration of the old model above is 0.25 m s^{-2}, because its speed increased by 0.25 m s^{-1} every second.

1.3.1 Performance tests

A car maker wants to compare the performance of a new model with the model being replaced. To do this, the velocity of each car is measured on a test track. Each vehicle accelerates as fast as possible to top velocity from a standing start. The results are listed in Table 1.1 below.

Table 1.1 *Performance of two models of a car*

Time from a standing start/s	0	20	40	60	80	100
Velocity of old model/m s^{-1}	0	5	10	15	20	20
Velocity of new model/m s^{-1}	0	6	12	18	18	18

Which car accelerates faster?

- The old model took 80 s to reach its top velocity of 20 m s^{-1}. Its velocity must have increased by 0.25 m s^{-1} every second for 80 seconds to reach its top velocity.
- The new model took 60 s to reach its top velocity of 18 m s^{-1}. Its velocity must have increased by 0.30 m s^{-1} every second for 60 s to reach its top velocity.

Clearly, the new model speeds up at a faster rate than the old model. In other words, its acceleration is greater.

> **Acceleration is defined as change of velocity per unit time.**

- The unit of acceleration is the **metre per second per second (m s^{-2})**.
- Acceleration is a **vector** quantity.
- **Deceleration** values are negative and signify that the velocity decreases with respect to time.

For a moving object that does not change direction, its acceleration at any point can be worked out from its rate of change of speed because there is no change of direction.

1.3.2 Uniform acceleration

Uniform acceleration is where the velocity of an object moving along a straight line changes at a constant rate. In other words, the acceleration is constant. Consider an object that accelerates uniformly from velocity, u, to velocity, v, in time, t, along a straight line. Figure 1.22 shows how its velocity changes with time.

$$\textbf{Acceleration, } a\textbf{, of the object} = \frac{\textbf{change of velocity}}{\textbf{time taken}} = \frac{(v - u)}{t}$$

$$a = \frac{v - u}{t}$$

where u is the initial velocity of the object and v is the velocity of the object at time t.

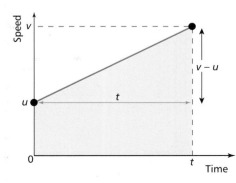

Fig 1.22 *Uniform acceleration*
Notice that the gradient of the line represents the acceleration $\left(= \dfrac{v - u}{t} \right)$

To calculate the velocity v at time t, rearranging this equation gives:

$$at = (v - u)$$

$$\therefore \qquad v = u + at$$

Worked example

The driver of a vehicle travelling at 8 m s^{-1} applies the brakes for 30 s and reduces the velocity of the vehicle to 2 m s^{-1}. Calculate the deceleration of the vehicle during this time.

Solution

$$u = 8 \text{ m s}^{-1}; v = 2 \text{ m s}^{-1}; t = 30 \text{ s}$$

$\therefore$
$$a = \frac{(v - u)}{t} = \frac{(2 - 8)}{30} = \frac{-6}{30} = -0.2 \text{ m s}^{-2}$$

1.3.3 Non-uniform acceleration

Non-uniform acceleration is where the direction of motion of an object changes, or its speed changes, at a varying rate. Figure 1.23 shows how the speed of an object increases for an object moving along a straight line with an increasing acceleration. This can be seen directly from the graph, because the gradient increases with time (i.e. the graph becomes steeper and steeper). The **gradient** represents the acceleration.

The acceleration at any point is the gradient of the tangent to the curve at that point. In Figure 1.23:

Gradient of tangent at point P = $\dfrac{\text{height of gradient triangle}}{\text{base of gradient triangle}}$

$$= \frac{4.0 - 1.0 \text{ m s}^{-1}}{2.0 \text{ s}} = 1.5 \text{ m s}^{-2}$$

Therefore, the acceleration at P is 1.5 m s^{-2}.

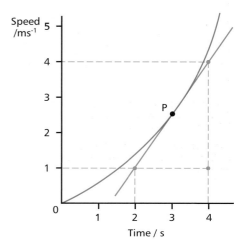

Fig 1.23 *Non-uniform acceleration*

> **Acceleration = gradient of the line on the velocity-time graph**

QUESTIONS

1 **a** An aeroplane taking off accelerates uniformly on a runway from a velocity of 4 m s^{-1} to a velocity of 64 m s^{-1} in 40 s. Calculate its acceleration.

   b A car travelling at a velocity of 20 m s^{-1} brakes sharply to a standstill in 8.0 seconds. Calculate its deceleration, assuming its velocity decreases uniformly.

2 A cyclist accelerates uniformly from a velocity of 2.5 m s^{-1} to a velocity of 7.0 m s^{-1} in a time of 10 s. Calculate:

   a the acceleration,

   b the velocity 2.0 s later, if she continued to accelerate at the same rate.

3 A train, on a straight journey between two stations, accelerates uniformly from rest for 20 s to a velocity of 12 m s^{-1}. It then travelled at a constant velocity for a further 40 s, before decelerating uniformly to rest in 30 s.

   a Sketch a velocity-time graph to represent its journey.

   b Calculate its acceleration in each part of the journey.

4 The velocity of an object released in water increased as shown in Figure 1.24.

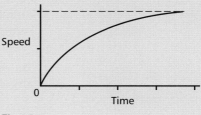

Fig 1.24

Describe how:

   a the velocity of the object changed with time,

   b the acceleration of the object changed with time.

Motion along a straight line at constant acceleration

1.4.1 The dynamics equations for constant acceleration

Consider an object that accelerates uniformly from initial speed, u, to final speed, v, in time t without change of direction. Figure 1.22, p.14, shows how its speed changes with time.

1. The acceleration $a = \dfrac{(v - u)}{t}$ (as explained in section 1.3.2)

   As before, rearranging this equation gives:

   $$v = u + at \qquad \text{(Equation 1)}$$

2. The distance moved, s = average speed × time taken

   Because the acceleration is uniform, the average speed $= \dfrac{(u + v)}{2}$

   therefore $\qquad\qquad\qquad s = \dfrac{(u + v)t}{2} \qquad$ (Equation 2)

3. By combining the two equations above, to eliminate v, a further useful equation is produced.

   Substitute $u + at$ in place of v in Equation 2. This gives:

   $$s = \frac{(u + (u + at))t}{2} = \frac{(u + u + at)t}{2} = \frac{(2ut + at^2)}{2}$$

   therefore $\qquad\qquad\qquad s = ut + \tfrac{1}{2}at^2 \qquad$ (Equation 3)

4. A fourth useful equation is obtained by combining Equations 1 and 2 to eliminate t. This can be done by multiplying:

   $$a = \frac{(v - u)}{t} \text{ and } s = \frac{(u + v)t}{2} \text{ together to give}$$

   $$as = \frac{(v - u)}{t} \times \frac{(u + v)t}{2}$$

   which simplifies to become:

   $$as = \frac{(v - u)(v + u)}{2} = \frac{(v^2 - uv + uv - u^2)}{2} = \frac{(v^2 - u^2)}{2}$$

   Therefore $2\,as = v^2 - u^2$

   Rearranging this equation gives:

   $$v^2 = u^2 + 2as \qquad \text{(Equation 4)}$$

The four equations, often referred to as the 'suvat' equations, are invaluable in any situation where the acceleration is constant.

1.4.2 Using a speed–time graph to find the distance moved

An object moving at constant speed

The distance moved in time t, s = speed × time taken = $v\,t$. This distance is represented on the graph by the **area under the line between the start and time t**. This is a rectangle of height corresponding to speed v and of base corresponding to the time t (Fig 1.25a).

WORKED EXAMPLE

A driver of a vehicle travelling at a speed of 30 m s⁻¹ on a motorway brakes sharply to a standstill in a distance of 100 m. Calculate the deceleration of the vehicle.

$u = 30$ m s⁻¹, $v = 0$, $s = 100$ m, $a = ?$

Solution
To find a, use $v^2 = u^2 + 2as$

Therefore $0 = u^2 + 2as$, because $v = 0$

Rearranging this equation gives:

$2as = -u^2$

$a = \dfrac{-u^2}{2s} = \dfrac{-30^2}{2 \times 100} = -4.5$ m s⁻²

Fig 1.25   a Constant speed

An object moving at constant acceleration

From section 1.4.1, the distance moved in time t, $s = \dfrac{(u + v)t}{2}$.

This distance is represented on the graph by the **area under the line between the start and time t**. This is a trapezium which has a base corresponding to

time t and an average height corresponding to the average speed $\dfrac{(u + v)t}{2}$.

Therefore the area of the trapezium ($=$ average height $\times$ base) corresponds to

$\dfrac{(u + v)t}{2}$

which is the distance moved (Fig 1.25b).

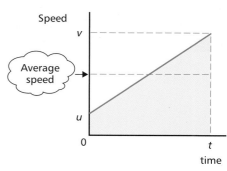

Fig 1.25   **b** Constant acceleration

An object moving with a changing acceleration

Let v represent the speed at time t and $v + \delta v$ represent the speed a short time later at $t + \delta t$ (δ is pronounced 'delta' from Greek).

Because the speed change δv is small compared with the speed v (Fig 1.25c), the distance travelled δs in the short time interval δt is given by $\delta s = v\delta t$.

This is represented on the graph by the **area of the shaded strip under the line which has a base corresponding to δt and a height corresponding to v**. In other words, $\delta s = v\,\delta t$ is represented by the area of this strip.

By considering the whole area under the line in strips of similar width, the total distance travelled from the start to time t_1 is therefore represented by the sum of the area of every strip: which is therefore the **total area** under the line.

Whatever the shape of the line of a speed-time graph:

> **Distance travelled = area under the line of a speed-time graph**

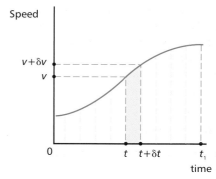

Fig 1.25   **c** Changing acceleration

QUESTIONS

1 A vehicle accelerates uniformly along a straight road, increasing its speed from 4.0 m s^{-1} to 30.0 m s^{-1} in 13 s. Calculate:

a its acceleration,

b the distance it moves in this time.

2 An aircraft lands on a runway at a speed of 40 m s^{-1} and brakes to a halt in a distance of 860 m. Calculate: **a** the braking time, **b** the deceleration of the aircraft.

3 A cyclist accelerates uniformly from rest to a speed of 6.0 m s^{-1} in 30 s, then brakes at uniform deceleration to a halt in a distance of 24 m.

a For the first part of the journey, calculate: (i) the acceleration, (ii) the distance travelled.

b For the second part of the journey, calculate: (i) the deceleration, (ii) the time taken.

c Sketch a speed-time graph for this journey,

d Use the graph to determine the average speed for the journey.

4 The speed of an athlete for the first 5 s of a sprint race is shown in Figure 1.26. Use the graph to determine:

a the initial acceleration of the athlete,

b the distance moved in the first 2 s,

c the distance moved in the next 2 s,

d the average speed over the first 4 s.

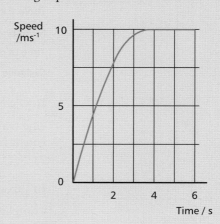

Fig 1.26

Free fall

1.5.1 Experimental tests

Does a heavy object fall faster than a lighter object?
Release a stone and a small coin at the same time. Which one hits the ground first? Galileo first discovered the answer to this question about four centuries ago. He reasoned that because any number of identical objects must fall at the same rate, then any one such object must fall at the same rate as the rest put together. So he concluded that any two objects must fall at the same rate, regardless of their relative weights. He was reported to have demonstrated the correctness of his theory by releasing two different weights from the top of the leaning Tower of Pisa.

The inclined plane test
Galileo wanted to know if a falling object speeds up as it falls. Clocks and stopwatches were devices of the future. The simplest test he could think of was to time a ball as it rolled down a plank. He devised a 'dripping water' clock, counting the volume of the drips as a measure of time. He measured how long the ball took to travel equal distances down the slope from rest. His measurements showed that the ball gained speed as it travelled down the slope. In other words, he showed that the ball accelerated as it rolled down the slope. He reasoned that the acceleration would be greater the steeper the slope. So he concluded that an object falling vertically accelerates.

1.5.2 Acceleration due to gravity

One way to investigate the free fall of a ball is to make a multi-flash photo or video clip of the ball's flight as it falls after being released from rest. To do this, a camera is used to record the ball's descent in a dark room with a flashing light switched on. The flashing light needs to flash at a known constant rate of about 20 flashes per second. A vertical metre rule can be used to provide a scale. Figure 1.27 shows a possible arrangement using a steel ball and a multi-flash photo taken with this arrangement.

For each image of the ball on the photograph, the time of descent of the ball and the distance fallen by the ball from rest can be measured directly. The photograph shows the ball speeds up as it falls, because it travels further between successive images. Measurements from this photograph are given in Table 1.2 below.

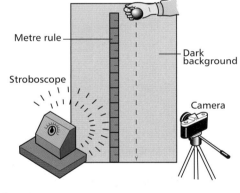

Metre rule

Stroboscope

Dark background

Camera

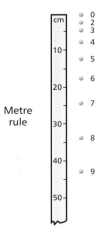

Metre rule

b *Freefall*

Fig 1.27 *Investigating free fall*

Table 1.2 A free-falling ball

Number of flashes after start	0	2	3	4	5	6	7	8	9
Time of descent, t/s	0	0.06	0.10	0.13	0.16	0.19	0.23	0.26	0.29
Distance fallen, s/m	0	0.02	0.04	0.07	0.12	0.17	0.24	0.33	0.42

How can we tell if the acceleration is constant from these results?

One way is to consider how the distance fallen, s, would vary with time, t, for constant acceleration. From section 1.4, we know that:

$$s = ut + \tfrac{1}{2}at^2, \text{ where } u = \text{the initial speed, and } a = \text{acceleration.}$$

In this experiment, $u = 0$ therefore $s = \tfrac{1}{2}at^2$ for constant acceleration, a.

Compare this equation with the general equation for a straight-line graph $y = mx + c$, where m is the gradient and c is the y-intercept. If we let y represent s and let x represent t^2, then $m = \tfrac{1}{2}a$ and $c = 0$. See 12.5.2 if necessary.

So a graph of s against t^2 should give a straight line through the origin. In addition, the gradient of the line $(=\frac{1}{2}a)$ can be measured and the acceleration $(=2\times \text{gradient})$ calculated (Fig 1.28).

As you can see, the graph is a straight line through the origin. We can therefore conclude that the equation $s=\frac{1}{2}at^2$ applies here, so the acceleration of a falling object is constant. Show for yourself that the gradient of the line is 5.0 m s^{-2} ($\pm$ 0.2 m s^{-2}), giving an acceleration of 10 m s^{-2}. Because there are no external forces acting on the object, apart from its weight, this value of acceleration is known as the **acceleration of free fall** and is represented by the symbol, **g**. Accurate measurements give a value of 9.8 m s^{-2} near the Earth's surface.

The 'suvat' equations on p.16 may be applied to any 'free fall' situation, where air resistance is negligible.

The equations can also be applied to situations where objects are thrown vertically upwards. As a general rule, apply the direction code + **for upwards and** − **for downwards** when values are inserted into the 'suvat' equations.

Worked example

$g = 9.8$ m s^{-2}

1 A coin was released from rest at the top of a well. It took 1.6 s to hit the bottom of the well. Calculate: **a** the distance fallen by the coin, **b** its speed just before impact.

Solution

$u = 0$, $t = 1.6$ s, $a = -9.8$ m s^{-2} ($-$ as g acts downwards)

a To find s, use $s = \frac{1}{2}at^2$ as $u = 0$
Therefore, $s = \frac{1}{2} \times -9.8 \times 1.6^2 = -12.5$ m ($-$ indicates 12.5 m downwards)

b To find v, use $v = u + at = 0 + (-9.8 \times 1.6) = -15.7$ m s^{-1}
($-$ indicates downward velocity)

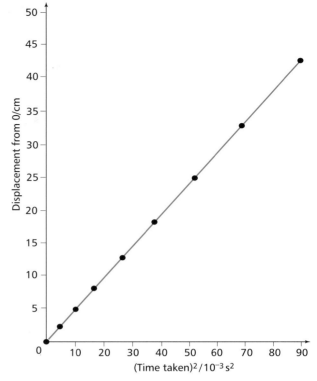

Fig 1.28 *A graph of s against t^2*

QUESTIONS

$g = 9.8$ m s^{-2}

1 A pebble, released at rest from a canal bridge, took 0.9 s to hit the water.
Calculate:

a the distance it fell before hitting the water,

b its speed just before hitting the water.

2 A spanner was dropped from a hot air balloon, when the balloon was at rest 50 m above the ground.
Calculate:

a the time taken for the spanner to hit the ground,

b the speed of impact of the spanner on hitting the ground.

3 A bungee jumper jumped off a platform 75 m above a lake, releasing a small object at the instant she jumped off the platform.

a Calculate (i) the time taken by the object to fall to the lake, (ii) the speed of impact of the object on hitting the water, assuming air resistance is negligible.

b Explain why the bungee jumper would take longer to descend than the time taken in part **a**(i).

4 An astronaut on the Moon threw an object 4.0 m vertically upwards and caught it again 4.5 s later.
Calculate:

a the acceleration due to gravity on the Moon,

b the speed of projection of the object,

c how high the object would have risen on the Earth, for the same speed of projection.

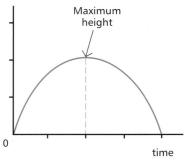

Fig 1.29 *Displacement against time for an object projected upwards*

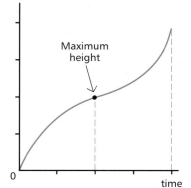

Fig 1.30 *Distance against time for an object projected upwards*

Fig 1.31 *Velocity-time graph*

1.6.1 The difference between a distance-time graph and a displacement-time graph

Displacement is distance in a given direction from a certain point. Consider a ball thrown directly upwards and caught when it returns. Its displacement from the instant it leaves the thrower:

- increases until it reaches maximum height,
- decreases to zero as it returns to the thrower from maximum height.

For example, if the ball rises to a maximum height of 2.0 m, on returning to the thrower its displacement from its initial position is zero. However, the distance it has travelled is 4.0 m.

The displacement of the object changes with time as shown by Figure 1.29. The line in this graph fits the equation $s = ut - \frac{1}{2}gt^2$, where s is the displacement, t is the time taken and u is the initial velocity of the object.

The **gradient** of the line represents the **velocity** of the object.

- Initially, the velocity is positive and large so the gradient is positive and large.
- As the ball rises, its velocity decreases so the gradient decreases.
- At maximum height, the velocity is zero so the gradient is zero.
- As the ball returns, the velocity becomes increasingly negative, corresponding to increasing speed in a downwards direction. So the gradient becomes increasingly negative.

The distance travelled by the object changes with time as shown by Figure 1.30. The **gradient** of this line represents the **speed**.

- From projection to maximum height, the shape is exactly the same as in Figure 1.29.
- After maximum height, the distance continues to increase so the line curves up, not down.

1.6.2 More about velocity-time graphs

Velocity is speed in a given direction. Consider how the velocity of an object thrown into the air changes with time. The object's velocity:

- decreases from its initial positive (i.e. upwards) value to zero at maximum height,
- increases in the negative (i.e. downwards) direction as it falls.

Figure 1.31 shows how the velocity of the object changes with time.

- **The gradient of the line is constant and negative, equal to the acceleration of free fall, g.** The acceleration of the object is the same when it descends as when it ascends, so the gradient of the line is always equal to -9.8 m s^{-2}.
- **The area under the line represents the displacement of the object from its starting position.**
  - **a** The area between the **positive** section of the line and the time axis represents the **displacement** during the **ascent**.
  - **b** The area between the **negative** section of the line and the time axis represents the displacement during the **descent**.

    Taking the area for **a** as positive and the area for **b** as negative, the total area is zero. This corresponds to zero for the total displacement.

Worked example

$g = 9.8 \text{ m s}^{-2}$

A ball released from a height of 1.20 m above a concrete floor rebounds to a height of 0.82 m.

a Calculate: (i) its time of descent, (ii) the speed of the ball immediately before it hits the floor.

b Calculate: the speed of the ball immediately after it leaves the floor.

c Sketch a velocity-time graph for the ball. Assume the contact time is negligible compared with the time of descent or ascent.

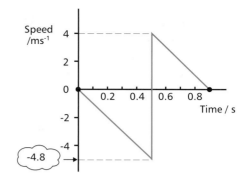

Fig 1.32

Solution

a $u = 0$, $a = -9.8 \text{ m s}^{-2}$, $s = -1.2 \text{ m}$

(i) To find t, use $s = ut + \frac{1}{2}at^2$

$$\therefore \quad -1.2 = 0 + 0.5 \times -9.8 \times t^2$$

$$-1.2 = -4.9t^2$$

$$t^2 = \frac{-1.2}{-4.9} = 0.245$$

$$t = 0.49 \text{ s}$$

(ii) To find v, use $v = u + at$

$$\therefore \quad v = 0 + -9.8 \times 0.49 = -4.8 \text{ m s}^{-1} \; (- \text{ for downwards})$$

b $v = 0$, $a = -9.8 \text{ m s}^{-2}$, $s = +0.82 \text{ m}$

To find u, use $v^2 = u^2 + 2as$

$$\therefore \quad 0 = u^2 + 2 \times -9.8 \times 0.82$$

$$u^2 = 15.7 \text{ m}^2 \text{ s}^{-2}$$

$$u = +4.0 \text{ m s}^{-1} \; (+ \text{ for upwards})$$

c See Figure 1.32. Note that the line has a constant gradient equal to the acceleration due to gravity, -9.8 m s^{-2}, except on impact.

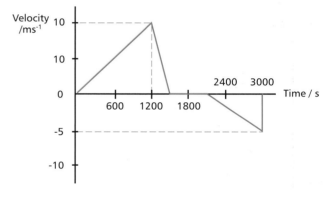

Fig 1.33

QUESTIONS

$g = 9.8 \text{ m s}^{-2}$

1 A swimmer swims 100 m from one end of a swimming pool to the other end at a constant speed of 1.2 m s^{-1}; then swims back at constant speed, returning to the starting point 210 s after starting.

a Calculate how long the swimmer takes to swim from: (i) the starting end to the other end, (ii) back to the start from the other end.

b For the swim from start to finish, sketch: (i) a displacement-time graph, (ii) a distance-time graph.

c Sketch a velocity-time graph for the swim.

2 A motor-cyclist travelling along a straight road at a constant speed of 8.8 m s^{-1} passes a cyclist travelling in the same direction at a speed of 2.2 m s^{-1}. After 200 s, the motor-cyclist stops.

a Calculate how long the motor-cyclist has to wait before the cyclist catches up.

b On the same axes, sketch a velocity-time graph for: (i) the motor-cyclist, (ii) the cyclist.

3 The graph (Fig 1.33) shows the velocity of a train on a straight track for 50 min after it left a station.

a (i) Describe how the displacement of the train from the station changed with time.

(ii) Sketch a graph to show how the displacement in part (i) varied with time.

b (i) Calculate how far from the station the train was after 50 min.

(ii) Calculate the total distance travelled by the train in this time.

4 A ball is released from a height of 1.8 m above a level surface and rebounds to a height of 0.90 m.

a Given $g = 9.8 \text{ m s}^{-2}$, calculate: (i) the duration of its descent, (ii) its velocity just before impact, (iii) the duration of its ascent, (iv) its velocity just after impact.

b Sketch a graph to show how its velocity changes with time from release to rebound at maximum height.

c Sketch a further graph to show how the displacement of the object changes with time.

More calculations on motion along a straight line

1.7.1 Motion along a straight line at constant acceleration

- The 'suvat' equations for motion at constant acceleration, a, are:

$$v = u + at \qquad \text{(Equation 1)}$$

$$s = \frac{(u + v)t}{2} \qquad \text{(Equation 2)}$$

$$s = ut + \tfrac{1}{2}at^2 \qquad \text{(Equation 3)}$$

$$v^2 = u^2 + 2as \qquad \text{(Equation 4)}$$

where s is the displacement in time t, u is the initial velocity and v is the final velocity.

- For motion along a straight line at constant acceleration, one direction along the line is 'positive' and the other direction is negative.

Worked example

A space vehicle moving towards a docking station, at a speed of 2.5 m s^{-1}, is 26 m from the docking station when its reverse thrust motors are switched on (to slow it down and stop it when it reaches the station). The vehicle decelerates uniformly until it comes to rest at the docking station, when its motors are switched off.

Calculate: **a** its deceleration, **b** how long it takes to stop, **c** its velocity, if its motors remained on for 5.0 s longer than necessary.

Solution

Let the + direction represent motion towards the docking station and − away from the station.

a and **b** Initial velocity, $u = +2.5$ m s^{-1}, final velocity, $v = 0$, displacement, $s = +26$ m.

To find its deceleration, a, use $\quad v^2 = u^2 + 2as$.

$$0 = 2.5^2 + 2a \times 26$$

$$-52\,a = 2.5^2$$

$$a = -\frac{2.5^2}{52} = -0.12 \text{ m s}^{-2}$$

To find the time taken, use $\quad v = u + at$.

$$0 = 2.5 - 0.12t$$

$$0.12t = 2.5$$

$$t = \frac{2.5}{0.12} = 21 \text{ s}$$

c Initial velocity, $u = 2.5$ m s^{-1}, acceleration, $a = -0.12$ m s^{-2}, time taken $t = 26$ s

To calculate its velocity v after 26 s, use $v = u + at = 2.5 - 0.12 \times 26$

$$= -0.62 \text{ m s}^{-1}$$

The velocity of the space vehicle would be 0.62 m s^{-1} away from the docking station after 26 s.

Fig 1.34 *A space vehicle docking*

1.7.2 Two-stage problems

Consider an object released from rest, falling then hitting a bed of sand. The motion is in two stages:

1. **Falling** motion due to gravity; acceleration $= g$ (downwards)
2. **Deceleration** in the sand; initial velocity = velocity of object just before impact.

The acceleration in each stage is **not** the same. The link between the two stages is that the velocity at the end of the first stage is the same as the velocity at the start of the second stage.

For example, consider a ball released from a height of 0.85 m above a bed of sand that creates an impression in the sand of depth 0.025 m. For directions, let + represent upwards and − represent downwards.

Stage 1
$u = 0$, $s = -0.85$ m, $a = -9.8$ m s^{-2}.

To calculate the speed of impact v, use $v^2 = u^2 + 2as$

∴ $$v^2 = 0^2 + 2 \times -9.8 \times -0.85 = 16.7$$
∴ $$v = -4.1 \text{ m s}^{-1}$$

Note $v^2 = 16.7$, so $v = -4.1$ or $+4.1$ m s^{-1}. The negative answer is chosen, as the ball is moving downwards.

Stage 2
$u = -4.1$ m s^{-1}, $v = 0$ (as the ball comes to rest in the sand), $s = -0.025$ m

To calculate the deceleration, a, use $v^2 = u^2 + 2as$.

∴ $$0^2 = (-4.1)^2 + 2a \times -0.025$$
∴ $$2a \times 0.025 = 16.7$$
$$a = \frac{16.7}{2 \times 0.025} = 334 \text{ m s}^{-2}$$

Note $a > 0$ and therefore in the opposite direction to the direction of motion, which is downwards. Thus the ball slows down in the sand with a deceleration of 334 m s^{-2}.

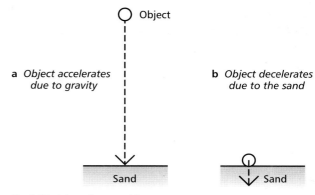

a *Object accelerates due to gravity* **b** *Object decelerates due to the sand*

Fig 1.35 *A two-stage problem*

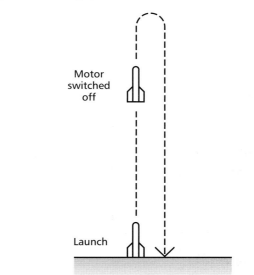

Fig 1.36

QUESTIONS

$g = 9.8$ m s^{-2}

1 A vehicle on a straight downhill road accelerated uniformly from a speed of 4.0 m s^{-1} to a speed of 29 m s^{-1} over a distance of 850 m, when the driver braked and stopped the vehicle in 28 s.

  a Calculate: (i) the time taken to reach 29 m s^{-1} from 4 m s^{-1}, (ii) its acceleration during this time.

  b Calculate: (i) the distance it travelled during deceleration, (ii) its average deceleration as it slowed down.

2 A rail wagon moving at a speed of 2.0 m s^{-1} on a level track reached a steady incline, which slowed it down in 15.0 s and caused it to reverse. Calculate:

  a the distance it moved up the incline,

  b its acceleration on the incline,

  c its velocity and position on the incline after 20.0 s.

3 A cyclist accelerated from rest at a constant acceleration of 0.4 m s^{-2} for 20 s, then stopped pedalling and slowed to a standstill at a constant deceleration over a distance of 260 m.

  a Calculate: (i) the distance travelled by the cyclist in the first 20 s, (ii) the speed of the cyclist at the end of this time.

  b Calculate: (i) the time taken to cover the distance of 260 m after she stopped pedalling, (ii) her deceleration during this time.

4 A rocket was launched directly upwards from rest. Its motors operated for 30 s after it left the launch pad, providing it with a constant vertical acceleration of 6.0 m s^{-2} during this time. Its motors then switched off (Fig 1.36).

  a Calculate: (i) its initial velocity, (ii) its height above the launch pad when its motors switched off.

  b Calculate its maximum height gain after its motors switched off.

  c Calculate the velocity with which it would hit the ground, if it fell from maximum height without the support of a parachute.

A **projectile** is any object acted upon only by the force of gravity. Three key principles apply to all projectiles:

- The **acceleration** of the object is always equal to g and acts downwards, because the force of gravity acts downwards. The acceleration therefore only affects the vertical motion of the object.
- The **horizontal velocity** of the object is constant, because the acceleration of the object does not have a horizontal component.
- The motions in the horizontal and vertical directions are **independent** of each other.

1.8.1 Vertical projection

If an object is projected vertically, it moves vertically as it has no horizontal motion. Its acceleration is 9.8 m s^{-2} downwards. Using the direction code ' + is upwards; − is downwards', its displacement, y, and velocity, v_y, after time t are given by:

$$v_y = u - gt$$
$$y = ut - \tfrac{1}{2}gt^2$$

where u is its initial velocity.

See section 1.7 for more about vertical projection.

1.8.2 Horizontal projection

A stone thrown from a cliff top follows a curved path downwards before it hits the water. If its initial projection was horizontal:

- Its path through the air becomes steeper and steeper as it drops.
- The faster it is projected, the further away it will fall into the sea.
- The time taken for it to fall into the sea does not depend on how fast it is projected.

Suppose two coins are released at the same time above a level floor, such that one coin drops vertically and the other is projected horizontally. Which one hits the floor first? In fact, they both hit the floor simultaneously. Try it! Why should the two coins hit the ground at the same time? They are both pulled to the ground by the force of gravity which gives each coin a downward acceleration, g. The coin that is projected horizontally experiences the same downward acceleration as the other coin. This downward acceleration does not affect the horizontal motion of the coin projected horizontally; only the vertical motion is affected.

Investigating horizontal projection

A stroboscope and a camera may be used to record the motion of a projectile. Figure 1.39 shows a multi-flash photograph of two balls, A and B, released at the same time. B was released from rest and dropped vertically; A was given an initial horizontal projection so it followed a curved path. The stroboscope flashed at a constant rate, so images of both balls were recorded at the same time.

- **The horizontal position** of A changes by equal distances between successive flashes. This shows that the horizontal component of A's velocity is constant.
- **The vertical position** of A and B changes at the same rate. At any instant, A is at the same level as B. This shows that A and B have the same vertical component of velocity at any instant.

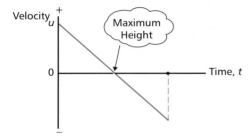

Fig 1.37 Upward projection

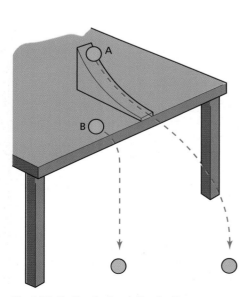

Fig 1.38 Testing horizontal projection

24

The projectile path of a ball projected horizontally

An object projected horizontally falls in an **arc** towards the ground. If its initial velocity is U, then at time t after projection(Fig 1.40):

- The **horizontal component** of its displacement,

$$x = Ut$$

  (because it moves horizontally at a constant speed)

- The **vertical component** of its displacement,

$$y = \tfrac{1}{2}gt^2$$

  (because it has no vertical component of its initial velocity)

- Its velocity has:

  a horizontal component $v_x = U$, and

  a vertical component $v_y = -gt$

  Note Its speed at time $t = \sqrt{(v_x{}^2 + v_y{}^2)}$

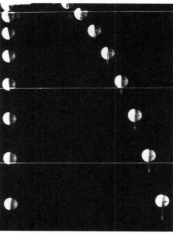

Fig 1.39 Multi-flash photo of two falling balls

Worked example

$g = 9.8 \text{ m s}^{-2}$

An object is projected horizontally at a speed of 15 m s^{-1} from the top of a tall tower (Fig 1.41) of height 35.0 m. Calculate:

a how long it takes to fall to the ground,

b how far it travels horizontally,

c its speed just before it hits the ground.

Solution

a $y = -35$ m, $a = -9.8 \text{ m s}^{-2}$ ($-$ for downwards)

$$y = \tfrac{1}{2}gt^2$$

$$\therefore \qquad t^2 = \frac{2y}{a} = \frac{2 \times -35}{-9.8} = 7.14 \text{ s}^2$$

$$\therefore \qquad t = 2.7 \text{ s}$$

b $U = 35 \text{ m s}^{-1}$, $t = 2.7$ s

$x = Ut = 35 \times 2.7 = 95$ m

c Just before impact, $v_x = U = 35 \text{ m s}^{-1}$ and $v_y = -gt = -9.8 \times 2.7 = 26.5 \text{ m s}^{-1}$

speed just before impact, $v = (v_x{}^2 + v_y{}^2)^{\frac{1}{2}} = (35^2 + 26.5^2)^{1/2} = 44 \text{ m s}^{-1}$

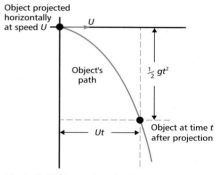

Fig 1.40 Horizontal projection

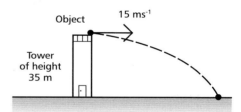

Fig 1.41

QUESTIONS

$g = 9.8 \text{ m s}^{-2}$

1 An object is released from a hot air balloon 50 m above the ground that is descending vertically at a speed of 4.0 m s^{-1}. Calculate:

   a the velocity of the object at the ground,

   b the duration of descent of the object,

   c the height of the balloon above the ground when the object hits the ground.

2 An object is projected horizontally at a speed of 16 m s^{-1} into the sea from a cliff top of height 45.0 m. Calculate:

   a how long it takes to reach the sea,

   b how far it travels horizontally,

   c its impact velocity.

3 A dart is thrown horizontally along a line which passes through the centre of a dartboard 2.3 m away from the point at which the dart was released. The dart hits the dartboard at a point 0.19 m below the centre. Calculate:

   a the time of flight of the dart,

   b its horizontal speed of projection.

4 A parcel is released from an aircraft travelling horizontally at a speed of 120 m s^{-1} above level ground. The parcel hits the ground 8.5 s later. Calculate:

   a the height of the aircraft above the ground,

   b the horizontal distance travelled in this time by (i) the parcel, (ii) the aircraft,

   c the speed of impact of the parcel at the ground.

Projectile motion 2

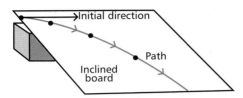

Fig 1.42 Using an inclined board

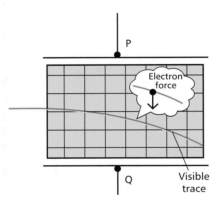

Fig 1.43 An electron beam on a parabolic path

(Vacuum tube used to contain apparatus not shown)

Fig 1.44 Projectile motion

1.9.1 Projectile-like motion

Any form of motion where an object experiences a constant acceleration in a different direction to its velocity will be like projectile motion. For example:

• The path of a ball rolling across an inclined board will be a projectile path. Figure 1.42 shows the idea. The object is projected across the top of the board from the side. Its path curves down the board and is **parabolic**. This is because the object is subjected to a constant acceleration acting down the board, and its initial velocity is across the board. The same equations as for projectile motion apply, with the motion down the board replacing the vertical motion.

• The path of a beam of electrons directed between two oppositely charged parallel plates is a **parabola**, as shown in Figure 1.43. Each electron in the beam is acted on by a constant force towards the positive plate, because the charge of an electron is negative. Therefore, each electron experiences a constant acceleration towards the positive plate. If its initial velocity is parallel to the plates, then its path is parabolic because its motion parallel to the plates is at **zero** acceleration; whereas its motion perpendicular to the plates is at **constant** (non-zero) acceleration.

1.9.2 Projection at angle θ and speed U above the horizontal

An object projected into the air follows a path that depends on its **speed** of projection and its **angle** of projection. In section 1.8, we looked at the special cases of vertical projection and of horizontal projection. In this section, we will consider the general case of projection at angle θ to the horizontal. The analysis can be applied to the motion of any object acted on by a force that is constant in **magnitude** and **direction**.

A **positive value** of θ will signify projection at angle θ **above** the horizontal.

A **negative value** of θ will signify projection at angle θ **below** the horizontal.

For an object projected at initial speed U at an angle θ to the horizontal, its **initial velocity** has:

• a **horizontal** component $u_x = U \cos \theta$
• a **vertical** component $u_y = U \sin \theta$

If necessary, see section 1.1 for resolving a vector into two perpendicular components.

Its **acceleration** has:

• a **horizontal** component $a_x = 0$
• a **vertical** component $a_y = -g$

For the **horizontal motion**, at time t after release:

• Its horizontal component of **velocity**, $v_x = U \cos \theta$ (unchanged), as its horizontal component of acceleration is zero.

• Its horizontal component of **displacement**, $x =$ its horizontal component of velocity × time taken $= Ut \cos \theta$

For the vertical motion,

• Its vertical component of **velocity**, $v_y = U \sin\theta - gt$
  This is obtained by applying the equation '$v = u + at$' to the vertical motion with $U \sin \theta$ for the initial speed u and $-g$ for the acceleration a.

• Its vertical component of **displacement**, $y = Ut \sin \theta - \frac{1}{2}gt^2$
  This is obtained by applying the equation '$s = ut + \frac{1}{2}at^2$' to the vertical motion with $U \sin \theta$ for the initial speed u and $-g$ for the acceleration a.

- Its speed at time $t = \sqrt{(v_x{}^2 + v_y{}^2)}$, in accordance with Pythagoras' rule for adding the two perpendicular components of a vector.

Worked example

$g = 9.8$ m s^{-2}

An arrow is fired at a speed of 48 m s^{-1} at an angle of 50° above the horizontal. Calculate:

a how long it takes to reach maximum height,

b its maximum height.

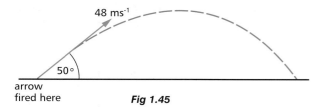

48 ms^{-1}

50°

arrow fired here

Fig 1.45

Solution

$U = 48$ m s^{-1}, $\theta = 50°$

a The vertical component of the initial velocity,
$u_y = U \sin \theta = 48 \sin 50 = 36.7$ m s^{-1}

The vertical component of velocity at maximum height,
$v_y = 0$

Using $v_y = U \sin \theta - gt$,

$0 = 36.7 - 9.8\,t$

$t = \dfrac{36.7}{9.8} = 3.74$ s $= 3.7$ s to 2 s.f.

b The maximum height can be calculated by using
$y = Ut \sin \theta - \frac{1}{2} gt^2$ with $t = 3.74$ s,

$\therefore$ maximum height $= 36.7 - 0.5 \times 9.8 \times 3.74^2$
$= 32$ m

QUESTIONS

$g = 9.8$ m s^{-2}

1 A ball was projected horizontally at a speed of 0.52 m s^{-1} across the top of an inclined board of width 600 mm and length 1200 mm (Fig 1.46). It reached the bottom of the board 0.90 s later.

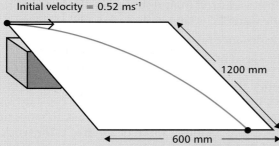

Initial velocity = 0.52 ms^{-1}

1200 mm

600 mm

Fig 1.46

Calculate:

a the distance travelled by the ball across the board,

b its acceleration on the board,

c its speed at the bottom of the board.

2 An arrow is fired at a speed of 45 m s^{-1} at an angle of 30° above the horizontal. Calculate:

a its maximum height,

b how long it takes to reach maximum height,

c how long it takes to return to the same horizontal level as it started at,

d the distance travelled horizontally to the point of return in **c**.

3 A cable car was travelling at a speed of 4.6 m s^{-1} in a direction that was 40° above the horizontal when an object was released from the cable car (Fig. 1.47).

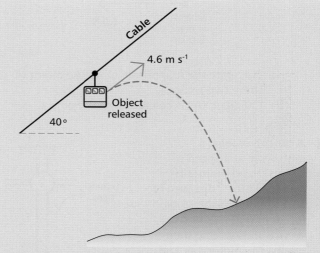

Cable

4.6 m s^{-1}

Object released

40°

Fig 1.47

a Calculate the horizontal and vertical components of velocity of the object at the instant it was released.

b The object took 5.8 s to fall to the ground below. Calculate: (i) the distance fallen by the object from the point of release, (ii) the horizontal distance travelled by the object from the point of release to where it hit the ground.

4 A cannon ball was fired from the top of a tower at an angle of elevation of 25°. The ball hit the ground 2.7 s later at a distance of 58 m away from the foot of the tower.

a Calculate the horizontal and vertical components of the velocity of the cannon ball at the instant it was fired.

b Calculate the height of the tower above the ground.

c Calculate the speed of impact of the cannon ball at the ground.

1 a (i) State the difference between a vector and a scalar quantity.

    (ii) Give one example of a vector quantity other than force.

    (iii) Give one example of a vector quantity other than speed.

b A 6.0 N force and a 5.0 N force act at right angles to each other on an object of mass 3.0 kg.

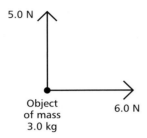

    (i) Calculate the magnitude of the resultant of these two forces.

    (ii) Calculate the acceleration of the object.

2 A car driver joins a motorway at a speed of 22 m s^{-1} and maintains this speed for a distance of 32.5 km, when she stops for 20 min at a motorway service station. She then rejoins the motorway and travels a further distance of 47.5 km in 40 min before leaving the motorway at a service station.

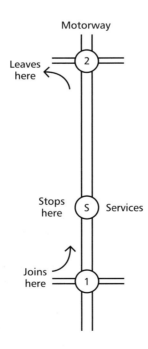

a Calculate the time taken for the first part of the journey.

b Sketch a speed-time graph for this journey on the motorway.

c Calculate the average speed of the car for the whole journey.

3 An aircraft taking off accelerated uniformly for 40 s from rest, before it left the ground after travelling a distance of 1600 m. Calculate:

a its average speed during take-off,

b its velocity at the instant it left the ground,

c its acceleration during take-off.

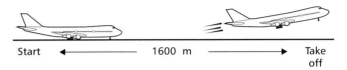

4 A train accelerated uniformly from rest for 40 s until its speed reached 15 m s^{-1}. It then travelled at a constant speed of 15 m s^{-1} for a further 60 s, before slowing down at constant deceleration and stopping 20 s later.

a Sketch a speed-time graph to represent its motion.

b Calculate its acceleration and the distance it travelled in each of the three stages of its motion.

c Calculate its average speed.

5 An object released from rest at the side of a river bridge hit the water 2.2 s later. Calculate:

a the distance fallen in air by the object,

b the speed of the object just before it hit the water.

6 A stationary railway truck on an inclined track was struck by a shunting engine moving at a constant speed of 0.6 m s^{-1}. The impact caused the truck to move up the incline at an initial speed of 2.0 m s^{-1}. The truck slowed down and stopped 10 s later.

a Calculate: (i) the distance travelled by the truck before it stopped, (ii) the magnitude and direction of its acceleration.

b The truck stopped for an instant, then rolled back down the track and hit the engine again. The engine continued to move up the track at 0.6 m s^{-1}.

    (i) Sketch a velocity-time graph on the same axes for the truck and the engine.

    (ii) Use your graph to show that the truck hit the engine a second time 14 s after it was first hit by the engine.

    (iii) Calculate the velocity of the truck immediately before this second impact.

7 A ball thrown vertically into the air left the thrower's hand when it was 1.6 m above the ground and hit the ground on return 3.1 s later.

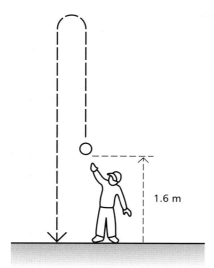

1.6 m

a Calculate: (i) the initial velocity of the ball, (ii) the maximum height of the ball above the ground.

b Calculate its velocity just before impact.

8 A dart thrown horizontally at a dartboard left the thrower's hand when it was 1.8 m from the dartboard and hit the dartboard 0.16 s later.

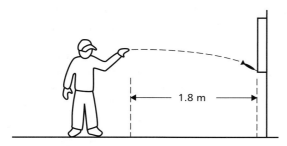

1.8 m

Calculate:

a the horizontal velocity of the dart,

b the distance fallen by the dart,

c the speed of the dart immediately before hitting the dartboard.

9 A cricketer on a level playing field threw a ball into the air at an angle of 30° above the horizontal. The ball was caught by another cricketer, 36 m away at the same level 1.8 s later.

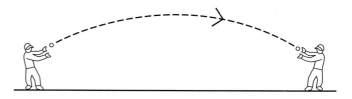

a Calculate the horizontal velocity of the ball.

b Show that the initial speed of the ball was 23 m s^{-1}.

c Calculate the maximum height reached by the ball.

10 A rocket launched vertically accelerated at a constant acceleration of 6.0 m s^{-2} for 25 s before its fuel supply ran out. It then rose to maximum height and fell back to the ground.

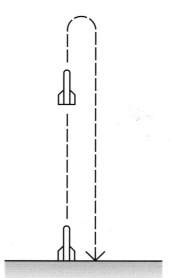

a (i) Calculate its velocity and its height when its fuel ran out.

(ii) Show that it reached a maximum height of 3.0 km.

b (i) Calculate how long it took to fall from maximum height to the ground.

(ii) Calculate its velocity immediately before impact with the ground.

(iii) Sketch a velocity-time graph for the motion of the rocket from launch to impact.

Forces in equilibrium

Balanced forces

2.1.1 Force as a vector

Force is a vector because it has **magnitude** and **direction**. The unit of force is the newton (N). As explained in section1.1:

- A force F can be resolved into two perpendicular components: $F \cos \theta$ and $F \sin \theta$ parallel and perpendicular to a line at angle θ to the line of action of the force.

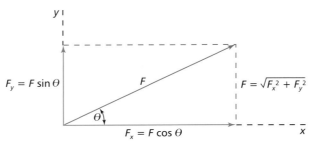

$F_y = F \sin \theta$

$F = \sqrt{F_x^2 + F_y^2}$

$F_x = F \cos \theta$

Fig 2.1 *Resolving a force*

- Two forces can be added together using a scale diagram or Pythagoras' theorem if the two forces are perpendicular to eath other. See Section 1.1 if necessary.

2.1.2 Equilibrium of a point object

When **two forces** act on a point object, the object is in equilibrium (i.e. at rest or moving at constant velocity) only if the two forces are equal and opposite to each other. The resultant of the two forces is therefore zero. The two forces are said to be **balanced**. For example, an object resting on a surface is acted on by its weight W (i.e. the force of gravity on it) acting downwards and a support force S from the surface acting upwards. Hence S is equal and opposite to W, provided the object is at rest (or moving at constant velocity) i.e. $S = W.$

When **three forces** act on a point object, their combined effect (i.e. resultant) is zero only if the resultant of any two of the forces is equal and opposite to the third force. One way of checking if the combined effect of the three forces is zero involves:

- resolving each force along the same parallel and perpendicular lines,
- balancing the components along each line.

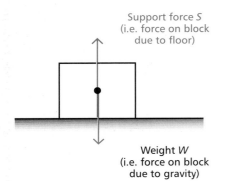

Support force S
(i.e. force on block due to floor)

Weight W
(i.e. force on block due to gravity)

Fig 2.2 *Balanced forces*

Example 1

A child of weight W on a swing is at rest, due to the swing seat being pulled to the side by a horizontal force F_1. The rope is then at an angle θ to the vertical, as shown in Figure 2.3.

Assuming the swing seat is of negligible weight, the swing seat is acted on by three forces: the weight of the child W, the horizontal force F_1 and the tension T in the rope. Resolving the tension T vertically and horizontally gives $T \cos \theta$ for the vertical component of T (which is upwards) and $T \sin \theta$ for the horizontal component of T. Therefore, the balance of forces:

- Horizontally: $F_1 = T \sin \theta$ • Vertically: $W = T \cos \theta$

Because $\sin^2 \theta + \cos^2 \theta = 1$ (see p 152), then $F_1^2 + W^2 = T^2 \sin^2 \theta + T^2 \cos^2 \theta = T^2$

$\therefore$
$$T^2 = F_1^2 + W^2$$

Also, because $\tan \theta = \dfrac{\sin \theta}{\cos \theta}$, then $\dfrac{F_1}{W} = \dfrac{T \sin \theta}{T \cos \theta} = \tan \theta$

$\therefore$ $$\tan \theta = \frac{F_1}{W}$$

Example 2

An object of weight W at rest on a rough slope is acted on by a frictional force F, which prevents it sliding down the slope, and a support force S from the slope perpendicular to the slope.

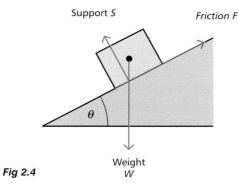

Fig 2.4

Resolving the three forces parallel and perpendicular to the slope gives:

- Horizontally: $F = W \sin \theta$ • Vertically: $S = W \cos \theta$

Because $\sin^2 \theta + \cos^2 \theta = 1$, then $F^2 + S^2 = W^2 \sin^2 \theta + W^2 \cos^2 \theta = W^2$

$\therefore$ $$W^2 = F^2 + S^2$$

Also, because $\tan \theta = \dfrac{\sin \theta}{\cos \theta}$, then $\dfrac{F}{S} = \dfrac{W \sin \theta}{W \cos \theta} = \tan \theta$

$\therefore$ $$\tan \theta = \frac{F}{S}$$

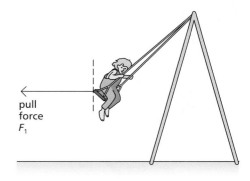

a On a swing

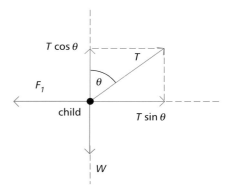

b Force diagram

Fig 2.3

QUESTIONS

1 A point object of weight 6.2 N is acted on by a horizontal force of 3.8 N:

 a Calculate the resultant of these two forces.

 b Determine the magnitude and direction of a third force acting on the object for it to be in equilibrium.

2 A small object of weight 5.4 N is at rest on a rough slope which is at an angle of 30° to the horizontal:

 a Sketch a diagram and show the three forces acting on the object.

 b Calculate: (i) the frictional force on the object, (ii) the support force from the slope on the object.

3 An archer pulled a bow string back until the two halves of the string were at 140° to each other (Fig 2.5). The force needed to hold the string in this position was 95 N.

 Calculate:

 a the tension in each part of the bow string in this position,

 b the resultant force on an arrow at the instant the bow string is released from this position.

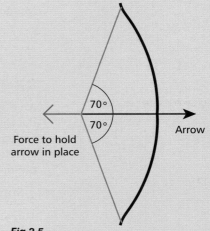

Fig 2.5

4 An elastic string is stretched horizontally between two fixed points, 0.80 m apart. An object of weight 4.0 N is suspended from the mid-point of the string, causing the mid-point to drop a distance of 0.12 m. Calculate:

 a the angle of each part of the string to the vertical,

 b the tension in each part of the string.

2.2

The Principle of Moments

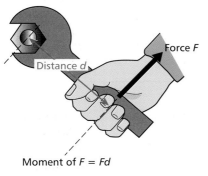

Moment of F = Fd

Fig 2.6 *A turning force*

2.2.1 Turning effects

Whenever you use a lever or a spanner, you are using a force to turn an object about a **pivot**. For example, if you use a spanner to loosen a wheel nut on a bicycle, you need to apply a force to the spanner to make it turn about the wheel axle. The effect of the force depends on how far it is applied from the wheel axle. The longer the spanner, the less force needed to loosen the nut. However, if the spanner is too long and the nut is too tight, the spanner could snap if too much force is applied to it.

> **The moment of a force about any point is the force × the perpendicular distance from the line of action of the force to the point. The unit of the moment of a force is the newton metre (N m).**

For a force F acting along a line of action at perpendicular distance d from a certain point:

$$\text{Moment of force} = Fd$$

Notes

- The greater the distance d, the greater the moment.
- The distance d is the **perpendicular** distance from the line of action of the force to the point (Fig 2.6).

Worked example

A spanner of length 0.24 m is used to tighten a wheel nut. The moment of the force must not exceed 60 N m, otherwise the wheel nut is damaged. Calculate the maximum force that should be applied to the spanner to tighten this nut.

Solution
Moment = force × distance

$$\therefore \qquad F \times 0.24 = 60$$
$$F = \frac{60}{0.24} = 250 \text{ N}$$

2.2.2 The Principle of Moments

An object that is not a point object is referred to as a **body**. Any such object turns if a force is applied to it anywhere other than through its centre of gravity. If a body is acted on by more than one force and it is in equilibrium, the turning effects of the forces must balance out. In more formal terms, considering the moments of the forces about any point, for equilibrium:

> **Sum of the clockwise moments = sum of the anticlockwise moments**

This statement is known as the **Principle of Moments.**

The balanced metre rule

1 Consider a uniform metre rule balanced on a pivot at its centre, supporting weights W_1 and W_2 suspended from the rule on either side of the pivot (Fig 2.7).

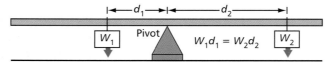

Fig 2.7 *The Principle of Moments*

- Weight W_1 provides an anticlockwise moment about the pivot of W_1d_1, where d_1 is the distance from the line of action of the weight to the pivot.
- Weight W_2 provides a clockwise moment about the pivot of W_2d_2, where d_2 is the distance from the line of action of the weight to the pivot.

For equilibrium, applying the Principle of Moments:

$$W_1d_1 = W_2d_2$$

2 If a third weight W_3 is suspended from the rule on the same side of the pivot as W_2 at distance d_3 from the pivot, then the rule can be rebalanced by increasing distance d_1.

At this new distance $d_1{}'$ for W_1:

$$W_1d_1{}' = W_2d_2 + W_3d_3$$

2.2.3 Centre of gravity

A tight-rope walker and a waiter serving drinks know just how important the centre of gravity of an object can be. One slight off-balance movement can be catastrophic.

The centre of gravity of a body is the point where the weight of the body may be considered to act.

Centre of gravity tests

- Support a ruler at its centre on the end of your finger. The centre of gravity of the ruler is directly above the point of support. Tip the ruler too much and it falls off, because the centre of gravity is no longer above the point of support.
- Balance a postcard on the end of a pencil. The centre of gravity of the postcard is directly above the point of support when the card is balanced. You should find that the centre of gravity of the card is at the mid-point of the card where the two diagonals cross.
- Find the centre of gravity of a triangular card. Figure 2.8 shows how to do this. It should be possible to balance the card at its centre of gravity on the end of a pencil.

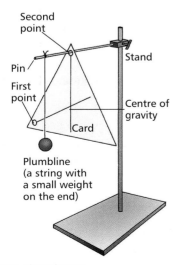

Fig 2.8 *A centre of gravity test*

Calculating the weight of a metre rule

- Locate the centre of gravity of a metre rule by balancing it horizontally on a horizontal knife edge. Note the position of the centre of gravity. The rule is said to be **uniform** if its centre of gravity is exactly at the middle of the rule.
- Balance the metre rule off-centre on a knife edge, using a known weight W_1 as shown in Figure 2.9. The position of the known weight needs to be adjusted gradually until the rule is exactly horizontal.

At this position:
- The known weight W_1 provides an anti-clockwise moment about the pivot of W_1d_1, where d_1 is the perpendicular distance from the line of action of W_1 to the pivot.
- The weight of the rule W_0 provides a clockwise moment of W_0d_0, where d_0 is the distance from the centre of gravity of the rule to the pivot.
- Applying the Principle of Moments:

$$W_0d_0 = W_1d_1$$

By measuring distance d_0 and d_1, the weight W_0 of the rule can therefore be calculated.

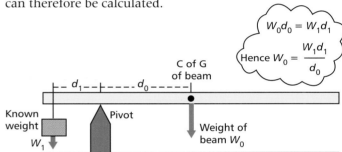

Fig 2.9 *Finding the weight of a beam*

QUESTIONS

1 A child of weight 200 N sits on a seesaw at a distance of 1.2 m from the pivot at the centre. The seesaw is balanced by a second child sitting on it at a distance of 0.8 m from the centre. Calculate the weight of the second child.

2 A metre rule pivoted at its centre of gravity supports a 3.0 N weight at its 5.0 cm mark, a 2.0 N weight at its 25 cm mark and a weight W at its 80 cm mark.

   a Sketch a diagram to represent this situation.

   b Calculate the weight W.

3 In question **2**, the 3.0 N weight and the 2.0 N weight are swapped with each other. Sketch the new arrangement and work out the new distance of weight W from the pivot.

4 A uniform metre rule supports a 4.5 N weight at its 100 mm mark. The rule is balanced horizontally on a horizontal knife edge at its 340 mm mark. Sketch the arrangement and calculate the weight of the rule.

More on moments

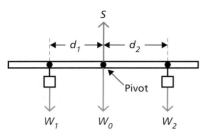

Fig 2.10 Support forces

2.3.1 Support forces

Single-support problems

When an object in equilibrium is supported at one point only, the support force on the object is equal and opposite to the total downward force acting on the object. For example, in Figure 2.10, a uniform rule is balanced on a knife edge at its centre of gravity with two additional weights W_1 and W_2 attached to the rule. The support force S on the rule from the knife edge must be equal to the total downward weight. Therefore:

$$S = W_1 + W_2 + W_0$$

where W_0 is the weight of the rule.

As explained in section 2.2, taking moments about the knife edge gives:

$$W_1 d_1 = W_2 d_2$$

Taking moments about a different point

Let's consider moments about the point where W_1 is attached to the rule:

Sum of the clockwise moments $= W_0 d_1 + W_2 (d_1 + d_2)$

Sum of the anticlockwise moments $= S d_1 = (W_1 + W_2 + W_0) d_1$

$$\therefore \qquad (W_1 + W_2 + W_0)\, d_1 = W_0 d_1 + W_2 (d_1 + d_2)$$

Multiplying out the brackets gives:

$$W_1 d_1 + W_2 d_1 + W_0\, d_1 = W_0 d_1 + W_2 d_1 + W_2 d_2$$

which simplifies to become: $W_1 d_1 = W_2 d_2$

This is the same as the equation obtained by taking moments about S. So moments can be taken about **any point**. It makes sense therefore to choose a point which one or more unknown forces act through, as such forces have **zero moment** about this point.

Two-support problems

Consider a uniform beam supported on two pillars X and Y, which are at distance D apart. The weight of the beam is 'shared' between the two pillars according to how far the beam's centre of gravity is from each pillar. For example:

- If the centre of gravity of the beam is mid-way between the pillars, the weight of the beam is 'shared' equally between the two pillars. In other words, the support force on the beam from each pillar is equal to half the weight of the beam.

- If the centre of gravity of the beam is at distance d_x from pillar X and distance d_y from pillar Y, as shown in Figure 2.11, then taking moments:

1 Where X is in contact with the beam:

$$S_y D = W d_x, \text{ where } S_y \text{ is the support force from pillar Y}$$

$$\text{so } \mathbf{S_y} = \frac{W d_x}{D}$$

2 Where Y is in contact with the beam:

$$S_x D = W d_y, \text{ where } S_x \text{ is the support force from pillar X}$$

$$\text{so } \mathbf{S_x} = \frac{W d_y}{D}$$

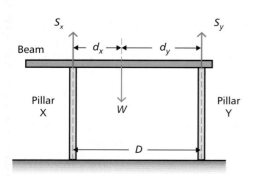

Fig 2.11 A two-support problem

Therefore, if the centre of gravity is closer to X than to Y, $d_x < d_y$ so $S_y < S_x$.

Worked example

A uniform beam of length 5.0 m and weight 120 N rests horizontally on the tops of two walls, X and Y (Fig 2.12), with its centre of gravity at a distance of 2.0 m from wall X and 1.5 m from wall Y. Calculate the support force on the beam from each wall.

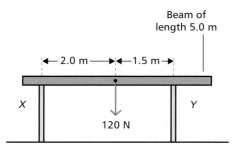

Fig 2.12

Solution

Let S_x and S_y represent the support forces at X and Y.

$$\therefore \qquad S_x + S_y = 120 \text{ N}$$

Taking moments about X gives:

Sum of the clockwise moments
  = weight of beam × distance from centre of gravity to X
  = 120 N × 2.0 m = 240 N m

Sum of the anticlockwise moments
  = support force S_y × distance from X to Y
  = S_y × 3.5 m

∴ Applying the Principle of Moments gives:

$$3.5\,S_y = 240$$

$$S_y = \frac{240}{3.5} = \textbf{69 N}$$

Therefore $\qquad S_x = W - S_y = 120 - 69 = \textbf{51 N}$

2.3.2 Couples

A **couple** is a pair of equal and opposite forces acting on a body, but not along the same line. Figure 2.13 shows a couple acting on a coil. The couple turns, or tries to turn, the coil.

The moment of a couple is referred to as a **torque:**

> **Torque of a couple = force × perpendicular distance between lines of action of the forces**

The total moment is the same, regardless of the point about which the moments are taken.

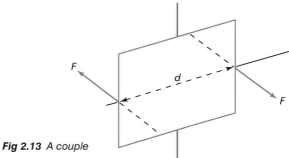

Fig 2.13 A couple

QUESTIONS

1 A metre rule of weight 1.2 N rests horizontally on two knife edges at the 100 mm mark and the 800 mm mark. Sketch the arrangement and calculate the support force on the rule due to each knife edge.

2 A uniform beam of weight 230 N and of length 10 m rests horizontally on the tops of two brick walls 8.5 m apart, such that a length of 1.0 m projects beyond one wall and 0.5 m projects beyond the other wall (Fig 2.14).

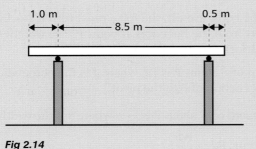

Fig 2.14

Calculate:

a the support force of each wall on the beam,

b the force of the beam on each wall.

3 A uniform bridge span of weight 1200 kN and of length 17.0 m rests on a support of width 1.0 m at either end. A stationary lorry of weight 60 kN is the only object on the bridge. Its centre of gravity is 3.0 m from the centre of the bridge (Fig 2.15). Calculate the support force on the bridge at each end.

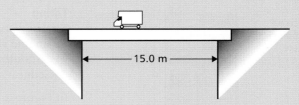

Fig 2.15

4 A uniform plank of weight 150 N and of length 4.0 m rests horizontally on two bricks. One of the bricks is at the end of the beam. The other brick is 1.0 m from the other end of the plank.

a Sketch the arrangement and calculate the support force on the plank from each brick.

b A child stands on the free end of the beam and just causes the other end to lift off its support.

Sketch this arrangement and calculate the weight of the child.

2.4.1 Stable and unstable equilibrium

Stable equilibrium

If a body in **stable equilibrium** is displaced then released, it returns to its equilbrium position. For example if an object such as a coat hanger hanging from a support is displaced slightly, it swings back to its equilibrium position.

Why does an object in stable equilbrium return to equilibrium when it is displaced then released?

- The reason is that the centre of gravity of the object is directly below the point of support when the object is at rest. The support force and the weight are directly equal and opposite to each other when the object is **in equilibrium**.

- However, when it is displaced then released, at the instant of release, the line of action of the weight no longer passes through the point of support so the weight **returns the object to equilibrium**.

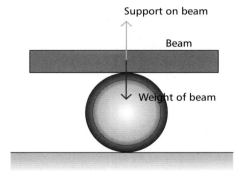

Support on beam

Beam

Weight of beam

Fig 2.16 *Unstable equilibrium*

Unstable equilibrium

A plank balanced on a drum is in **unstable equilibrium**. If it is displaced slightly from equilibrium then released, the plank will roll off the drum.

- The reason is that the centre of gravity of the plank is directly above the point of support when it is **in equilibrium**. The support force is exactly equal and opposite to the weight.

- If the plank is displaced slightly, the centre of gravity is no longer above the point of support. The weight therefore acts to turn the plank **further from the equilibrium position**.

2.4.2 Tilting and toppling

Skittles at a bowling alley are easy to knock over because they are top-heavy. This means that the centre of gravity is too high and the base is too narrow. A slight nudge from a ball causes a skittle to tilt then tip over.

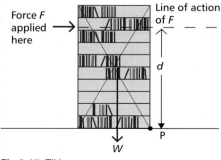

Force *F* applied here

Line of action of *F*

d

W

P

Fig 2.17 *Tilting*

Tilting

Tilting is where an object at rest on a surface is acted on by a force that raises it up on one side. For example, if a horizontal force is applied to the top of a tall free-standing bookcase, the force can make the bookcase tilt about its base along one edge.

In Figure 2.17, to make the bookcase tilt, the force must turn it clockwise about point P. The entire support from the floor acts at point P. The weight of the bookcase provides an anticlockwise moment about P.

- The **clockwise** moment of F about P is Fd, where d is the perpendicular distance from the line of action of F to the pivot.

- The **anticlockwise** moment of W about P is $\dfrac{Wb}{2}$, where b is the width of the base.

Therefore, for tilting to occur: $\boldsymbol{Fd > \dfrac{Wb}{2}}$

Toppling

A tilted object will **topple** over if it is tilted too far. For example, a tractor on a hill could topple over sideways if the hill is too steep. If an object on a flat surface is tilted more and more, the line of action of its weight (which is through its centre of gravity) passes closer and closer to the 'pivot'. If the object is tilted so much that the line of action of its weight passes beyond the pivot, the object will topple over if allowed to. The position where the line of action of the weight passes through the 'pivot' is the furthest it can be tilted without toppling. Beyond this position, it topples over if it is released.

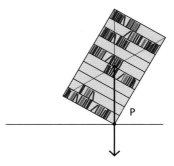

Fig 2.18 *Toppling over*

2.4.3 **On a slope**

A tall object on a slope will topple over if the slope is too great. For example, a high-sided vehicle on a road with a sideways slope will tilt over. If the slope is too great, the vehicle will topple over. This will happen if the line of action of the weight (passing through the centre of gravity of the object) lies outside the **wheel base** of the vehicle. In Figure 2.19, the vehicle will not topple over because the line of action of the weight lies within the wheel base.

Consider the forces acting on the vehicle on a slope when it is at rest. The sideways friction F, the support forces S_x and S_y and the force of gravity on the vehicle (i.e. its weight) act as shown in Figure 2.19. For equilibrium, resolving the forces parallel and perpendicular to the slope gives:

• **Parallel** to the slope:

$$F = W \sin \theta$$

• **Perpendicular** to the slope:

$$S_x + S_y = W \cos \theta$$

Note that S_x is greater than S_y because X is lower than Y.

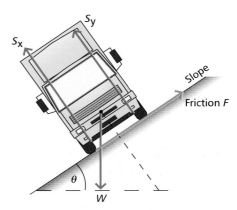

Fig 2.19 *On a slope*

QUESTIONS

1 Explain why a bookcase with books on its top shelf only is less stable than if the books were on the bottom shelf.

2 An empty wardrobe of weight 400 N has a square base 0.8 m × 0.8 m and a height of 1.8 m. A horizontal force is applied to the top edge of the wardrobe to make it tilt. Calculate the force needed to lift the wardrobe base off the floor along one side (Fig 2.20).

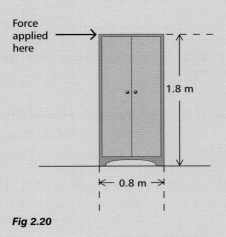

Fig 2.20

3 A vehicle has a wheel base of 1.8 m and a centre of gravity, when unloaded, which is 0.8 m from the ground (Fig 2.21).

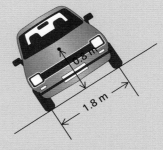

Fig 2.21

a The vehicle is tested for stability on an adjustable slope. Calculate the maximum angle of the slope to the horizontal if the vehicle is not to topple over.

b If the vehicle carries a full load of people, will it be more or less likely to topple over on a slope? Explain your answer.

4 Discuss whether or not a high-sided heavy goods lorry is more or less likely to be affected by strong side winds when it is fully loaded compared to when it is empty.

2.5.1 The triangle of forces

For a point object acted on by **three forces** to be in equilibrium, the three forces must give an overall **resultant of zero**. The three forces as vectors should form a **triangle**. In other words, for three forces F_1, F_2 and F_3 to give zero resultant:

$$\textbf{Vector sum, } \boldsymbol{F_1 + F_2 + F_3 = 0}$$

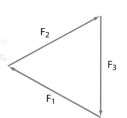

Fig 2.22 *Triangle of forces*

As explained in section 2.1, any two of the forces gives a resultant which is represented by the third side of the triangle. Therefore, for equilibrium, the third force must be represented by the third side of the triangle. For example, the resultant of $F_1 + F_2$ is equal and opposite to F_3:

i.e. $$\boldsymbol{F_1 + F_2 = - F_3}$$

The sine rule (see section 12.2) can be applied to the triangle of forces to find an unknown force or angle, given the other forces and angles in the triangle:

$$\frac{F_1}{\sin \theta_1} = \frac{F_2}{\sin \theta_2} = \frac{F_3}{\sin \theta_3}$$

where θ_1, θ_2 and θ_3 are the angles opposite sides F_1, F_2 and F_3 repectively.

The closed polygon

The triangle of forces rule can be extended for any number of forces acting on an object. If the object is in equilibrium, the force vectors drawn end-to-end must form a closed polygon. In other words, the tip of the last force vector must join the tail of the first force vector. Figure 2.23 shows the idea. Unfortunately, the sine rule can't be applied here; so the forces must be resolved in the same parallel and perpendicular directions to calculate an unknown force, given all the other forces.

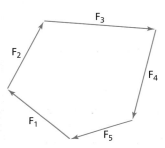

Fig 2.23 *The closed polygon*

2.5.2 The conditions for equilibrium of a body

Free-body force diagrams

When two objects interact, they always exert equal and opposite forces on one another. A diagram showing the forces acting on an object can become very complicated if it also shows the forces the object exerts on other objects. A **free-body force diagram** shows only the forces acting on the object.

Equilibrium

An object in equilibrium is either at rest or it moves with a constant velocity. In general, the forces acting on a body will not all act through the centre of gravity of the body. If the body is in equilibrium, the **turning effects** of the forces must balance out as well giving zero resultant.

For a body at rest:

• The **forces** must balance each other out (i.e. the resultant force must be zero).
• The Principle of Moments must apply (i.e. the **moments of the forces** about the same point must balance out).

Worked example

A uniform shelf of width 0.6 m and of weight 12 N is attached to a wall by hinges and is supported horizontally by two parallel cords attached at two of the corners of the shelf, as shown in Figure 2.24. The other end of each cord is fixed to the wall 0.4 m above the hinge. Calculate:

a the angle between each cord and the shelf,
b the tension in each cord.

Solution

a Let the angle between each cord and the shelf be θ.

From Figure 2.24, $\tan \theta = \dfrac{0.4}{0.6}$ so $\theta = 34°$

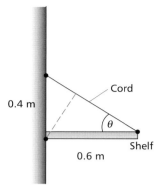

Fig 2.24

b Taking moments about the hinge gives:

- Sum of the clockwise moments
  = weight of shelf × distance from hinge to the centre of gravity of the shelf
  = $12 \times 0.3 = 3.6$ N m

- Sum of the anticlockwise moments = $2\,Td$, where t is the tension in each cord and d is the perpendicular distance from the hinge to either cord.

From Figure 2.24, it can be seen that $d = 0.6 \sin \theta = 0.6 \sin 34 = 0.34$ m

∴ Applying the Principle of Moments gives:

$$2 \times 0.34 \times T = 3.6$$
$$T = 5.3 \text{ N}$$

QUESTIONS

1 A uniform plank of length 5.0 m rests horizontally on two bricks, which are 0.5 m from either end. A child of weight 200 N stands on one end of the plank and causes the other end to lift, so it is no longer supported at that end.

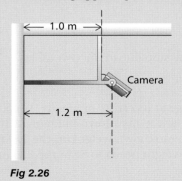

Fig 2.25

Calculate:

a the weight of the plank,

b the support force acting on the plank from the supporting brick.

2 A security camera is supported by a frame which is fixed to a wall and ceiling (Fig 2.26). The support structure must be strong enough to withstand the effect of a downward force of 1500 N acting on the camera, in case the camera is gripped by someone below it.

Fig 2.26

Calculate:

a the moment of a downward force of 1500 N on the camera about the point where the support structure is attached to the wall,

b the extra force of the vertical strut supporting the frame, when the camera is pulled with a downward force of 1500 N.

3 A crane is used to raise one end of a 15 kN girder of length 10.0 m off the ground. When the end of the girder is at rest 6.0 m off the ground, the crane cable is perpendicular to the girder (Fig 2.27). Calculate the tension in the cable.

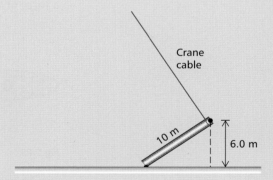

Fig 2.27

4 In question **3**, show that the support force on the girder from the ground has a horizontal component of 3.6 N and a vertical component of 10.2 N. Hence calculate the magnitude of the support force.

1 Calculate the magnitude of the resultant of a 6.0 N force and a 9.0 N force acting on a point object when the two forces act:

   a in the same direction,

   b in opposite directions,

   c at 90° to each other.

2 A point object in equilbrium is acted on by a 3 N force, a 6 N force and a 7 N force. What is the resultant force on the object if the 7 N force is removed?

3 A point object of weight 5.4 N in equilibrium is acted on by a horizontal force of 4.2 N and a second force F:

   a By considering the triangle of forces rule, determine the magnitude of F.

   b Calculate the angle between the direction of F and the horizontal.

4 An object of weight 7.5 N hangs on the end of a cord which is attached to the mid-point of a wire, stretched between two points on the same horizontal level. Each half of the wire is at 12° to the horizontal. Calculate the tension in each half of the wire.

Fig 2.28

5 A ship is towed at constant speed by two tug boats, each pulling the ship with a force of 9 kN. The angle between the tug-boat cables is 40°.

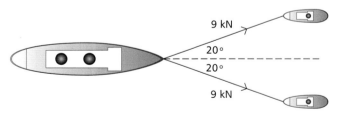

Fig 2.29

   a Calculate the resultant force on the ship due to the two cables.

   b Calculate the drag force on the ship.

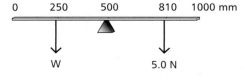

Fig 2.30

6 A metre rule is pivoted on a knife edge at its centre of gravity, supporting a weight of 5.0 N and an unknown weight W (as shown in Figure 2.30). To balance the rule horizontally with the unknown weight on the 250 mm mark of the rule, the position of the 5.0 N weight needs to be at the 810 mm mark.

   a Calculate the unknown weight.

   b Calculate the support force on the rule from the knife edge.

7 In Figure 2.30, a 2.5 N weight is also suspended from the rule at its 400 mm mark. What adjustment needs to be made to the position of the 5.0 N weight to rebalance the rule?

8 A uniform metre rule is balanced horizontally on a knife edge at its 350 mm mark by placing a 3.0 N weight on the rule at its 10 mm mark.

  a Sketch the arrangement and calculate the weight of the rule.

  b Calculate the support force on the rule from the knife edge.

9 A diving board has a length 4.0 m and a weight of 250 N. It is bolted to the ground at one end and projects by a length of 3.0 m beyond the edge of the swimming pool (Fig 2.31). A person of weight 650 N stands on the free end of the diving board. Calculate:

  a the force on the bolts,

  b the force on the edge of the swimming pool.

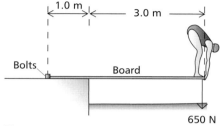

Fig 2.31

10 A uniform beam XY of weight 1200 N, and of length 5.0 m, is supported horizontally on a concrete pillar at each end. A person of weight 500 N sits on the beam, at a distance of 1.5 m from end X:

  a Sketch a free-body force diagram of the beam.

  b Calculate the support force on the beam from each pillar.

11 A bridge crane used at a freight depot consists of a horizontal span of length 12 m fixed at each end to a vertical pillar (Fig 2.32).

  a When the bridge crane supports a load of 380 kN at its centre, a force of 1600 kN is exerted on each pillar. Calculate the weight of the horizontal span.

  b The same load is moved across a distance of 2.0 m by the bridge crane. Sketch a free-body force diagram of the horizontal span, and calculate the force exerted on each pillar.

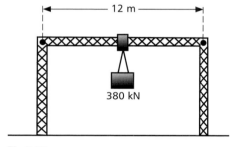

Fig 2.32

12 A curtain pole of weight 24 N and of length 3.2 m is supported horizontally by two wall-mounted supports X and Y, which are 0.8 m and 1.2 m from each end respectively.

  a Sketch the free-body force diagram for this arrangement, and calculate the force on each support when there are no curtains on the pole.

  b When the pole supports a pair of curtains of total weight 90 N drawn along the full length of the pole, what will be the force on each support?

13 A steel girder of weight 22 kN and of length 14 m is lifted off the ground at one end, by means of a crane. When the raised end is 2.0 m above the ground, the cable is vertical.

  a Sketch a free-body force diagram of the girder in this position.

  b Calculate the tension in the cable at this position and the force of the girder on the ground.

14 A rectangular picture, 0.80 m deep, 1.0 m wide and of weight 24 N, hangs on a wall supported by a cord attached to the frame at each of the top corners (Fig 2.33). Each section of the cord makes an angle of 25° with the picture, which is horizontal along its width.

  a Copy the diagram and mark the forces acting on the picture.

  b Calculate the tension in each section of the cord.

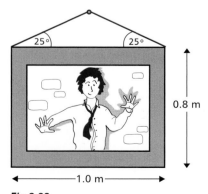

Fig 2.33

1 A tug boat is used to pull a barge along a river at constant velocity, as shown below. The tug boat pulls on the barge with a force of 4.5 kN. The tug-boat engine drives the tug boat forward with a force of 7.0 kN.

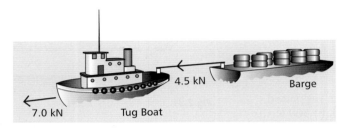

Calculate:

a the drag force on the tug boat due to the water,

b the drag force on the barge due to the water.

2 A yacht is moving due North as a result of a force, due to the wind, of 350 N in a horizontal direction of 40° East of due North, as shown below.

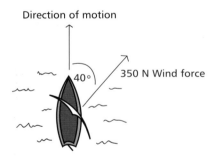

a Calculate the component of the force of the wind:
  (i) in the direction the yacht is moving,
  (ii) perpendicular to the direction in which the yacht is moving.

b Explain why the crew of the yacht need to lean out, as shown below, to prevent the yacht capsizing due to the force of the wind.

3 A car, of weight 6.2 kN, is towed at constant speed by a van on an uphill road as shown below. The tow rope is at an angle of 15° above the horizontal, when the road is 5° above the horizontal. The tension in the tow rope pulls the car with a force of 580 N.

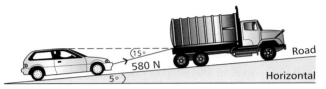

a Calculate the component of the tension in the tow rope:
  (i) parallel to the road,
  (ii) perpendicular to the road.

b Calculate the component of the car's weight:
  (i) parallel to the road,
  (ii) perpendicular to the road.

c By comparing your answers to parts **a**(i) and **b**(i), calculate the magnitude and direction of the force of friction acting on the car.

d The van's weight is 12 kN. By considering the component of the van's weight parallel to the road and the component of the tension in the tow rope, show that the force of the van's engine must be at least 1.6 kN.

4 A uniform girder, of weight 14 kN and of length 8.0 m, rests horizontally on two wooden blocks, one at 1.0 m from one end of the girder and the other at 0.5 m from the other end (as shown below). Calculate the force of the girder on each block.

5 a (i) Explain what is meant by the centre of gravity of a body.
  (ii) Define the moment of a force about a point.

b A builder needs to measure the weight of a bag of sand using a set of scales that can measure weights up to 200 N. He discovers that the bag weighs more than 200 N, so he devises the arrangement shown on page 43 to weigh the bag. This arrangement consists of a uniform plank, of length 4.5 m, resting horizontally with one end on a brick and the other end on the scales.

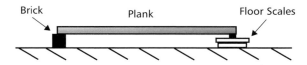

(a) Without the bag of sand

(i) Without the bag of sand on the plank, the scales read 30 N. Calculate the weight of the plank.

(ii) With the bag of sand on the plank, the scales read 200 N when the bag is 1.5 m from the end of the plank on the scales. Show that the bag of sand has a weight of 255 N.

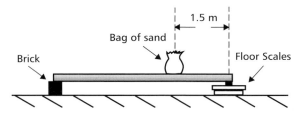

(b) With the bag of sand

6 A flower basket, of weight 35 N, is suspended on the lower end of a rope which is tied at its upper end to a horizontal bar. The force of a horizontal wind blows the basket to one side, causing the rope to make an angle of 20° with the vertical, as shown below.

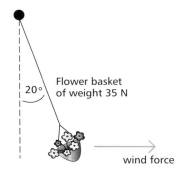

a Show that the tension in the rope in this position is 37 N.

b Calculate the force of the wind on the flower basket.

7 A paraglider, moving at constant velocity and constant height, is pulled by a cable which is at an angle of 35° to the horizontal. The force of the cable on the paraglider is 185 N.

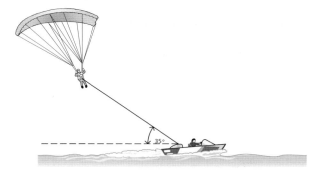

a Calculate the horizontal and vertical components of this force.

b With the aid of a force diagram, explain why the force of the parachute on the paraglider is not in the opposite direction to the force of the cable on the paraglider.

8 A bridge crane is used to transfer a container of weight 60 kN from a rail wagon onto a container lorry. The horizontal frame of the bridge crane has a length of 14 m and a total weight of 200 kN. It rests on two vertical steel pillars, as shown below.

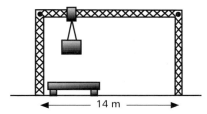

Calculate the force on each vertical pillar due to the weight of the horizontal frame and the container when:

a the container is suspended at the centre of the horizontal frame,

b the container is 4.0 m from the centre.

3 *Forces in action*

3.1 Force and acceleration

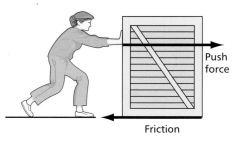

Fig 3.1 *Overcoming friction*

3.1.1 Motion without force

Motorists on icy roads in winter need to be very careful, because the tyres of a car have little or no 'grip' on the ice. Moving from a standstill on ice is very difficult. Stopping on ice is almost impossible, as a car moving on ice will slide when the brakes are applied. **Friction** is a hidden force that we don't usually think about until it is absent!

If you have ever tried to push a heavy crate across a rough concrete floor, you will know all about friction. The push force is opposed by friction and as soon as you stop pushing, friction stops the crate moving. If the crate had been pushed onto a patch of ice, it would have moved across the ice without any further push needed.

Figure 3.2 shows an air track which allows **motion** to be observed in the absence of friction. The glider on the air track floats on a cushion of air. Provided the track is level, the glider moves at a constant velocity along the track because friction is absent.

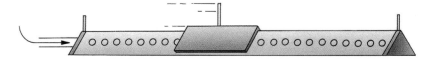

Fig 3.2 *The linear air track*

Newton's First Law of motion

> **Objects either stay at rest, or remain in uniform motion (i.e. at constant velocity), unless acted on by a force.**

Sir Isaac Newton was the first person to realise that a moving object stays moving at constant velocity unless acted on by a force. He recognised that when an object is acted on by a **resultant** force, the result is to change the object's velocity. In other words, an object moving at **constant** velocity is either:

- acted on by **no forces**, or
- the forces acting on it are **balanced** (i.e. the **resultant force is zero**).

3.1.2 Investigating force and motion

How does the velocity of an object change if it is acted on by a constant force? Figure 3.3 shows how this can be investigated using a dynamics trolley and a ticker-tape timer. The timer prints dots on the tape at a constant rate of 50 dots per second; so the faster the tape is moving, the greater the spacing between adjacent dots. An electronic timer linked to a computer can be used in place of the ticker-tape timer.

The trolley is pulled along a sloped runway by means of one or more elastic bands stretched to the same length. The runway is sloped just enough to compensate for friction. To test for the correct slope, the trolley should move down the runway at constant speed after being given a brief push.

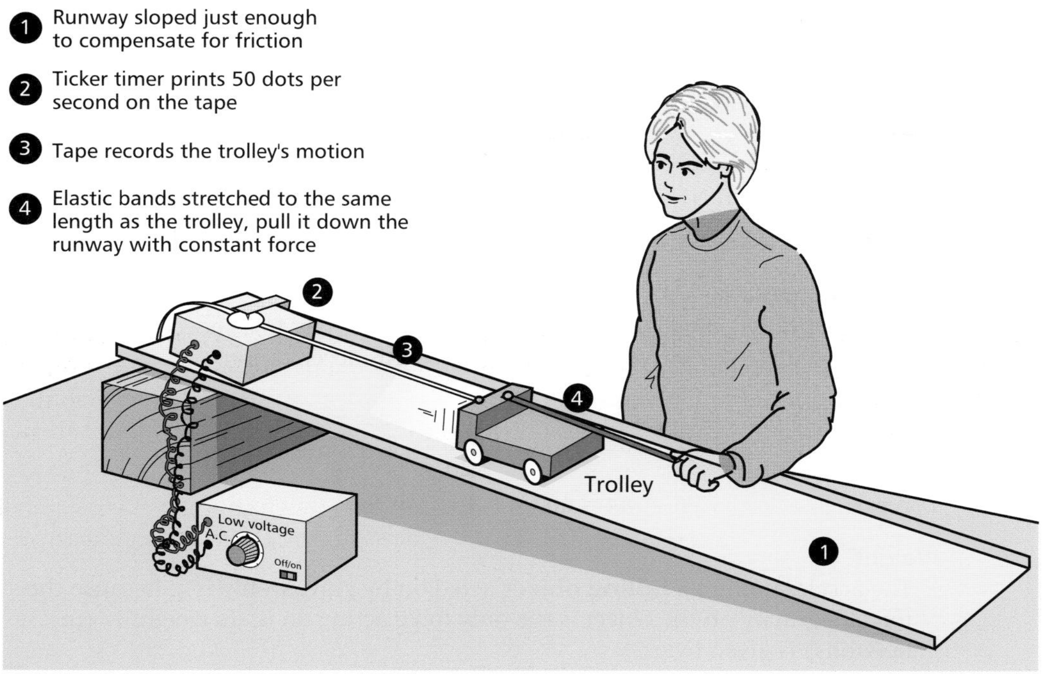

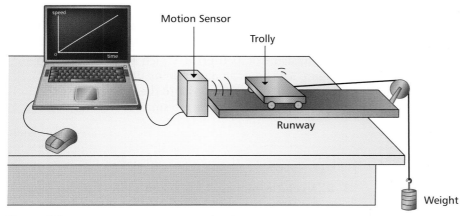

① Runway sloped just enough to compensate for friction

② Ticker timer prints 50 dots per second on the tape

③ Tape records the trolley's motion

④ Elastic bands stretched to the same length as the trolley, pull it down the runway with constant force

Fig 3.3 *Investigating force and motion*

- If a ticker-tape timer is used, the length of each 'ten dot' section of the tape gives a measure of the speed as the middle of that section went through the ticker timer. Cutting the tape into 'ten dot' sections enables a speed-time chart to be made, as shown in Figure 3.4.

- If an electronic timer linked to a computer is used, the measurements should be displayed directly as a speed-time graph (Fig 3.5).

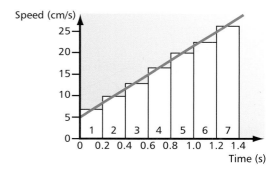

Fig 3.4 *Making a tape chart*

Fig 3.5 *Using a computer to measure acceleration*

The speed-time graph should show that the speed increased at a constant rate. The **acceleration** of the trolley is therefore constant, and can be measured from the speed-time graph. Table 3.1 shows typical measurements using different amounts of **force** (i.e. one, or two, or three elastic bands in 'parallel') and different amounts of **mass** (i.e. a single trolley, or a double trolley, or a triple trolley).

The results in Table 3.1 show that the force is proportional to the mass × the acceleration. In other words, if a force F acts on an object of mass m, the object undergoes an acceleration a such that:

$$F \text{ is proportional to } ma$$

Table 3.1 *Varying force and mass*

Force (i.e. no. of elastic bands)	1	2	3	1	2	3
Mass (i.e. no of trolleys)	1	1	1	2	2	2
Acceleration (m s⁻²)	12	24	36	6	12	18
Mass × acceleration	12	24	36	12	24	36

Newton's Second Law for constant mass

By defining the unit of force, the **Newton (N)**, as the amount of force that will give an object of mass 1 kg an acceleration of 1 m s^{-2}, the proportionality statement can be expressed as an equation:

$$F = ma$$

where F = force (in N), m = mass (in kg), a = acceleration (in m s^{-2}).

This equation is known as Newton's **Second Law** for constant mass.

Worked example

A vehicle of mass 600 kg accelerates uniformly from rest to a speed of 8.0 m s^{-2} in 20 s. Calculate the force needed to produce this acceleration.

Solution

Acceleration, $\quad a = (v - u)/t = \dfrac{(8.0 - 0)}{20} = 0.4 \text{ m s}^{-2}$

Force, $\quad F = ma = 600 \times 0.4 = 240 \text{ N}$

Weight

- The acceleration of a **falling object** acted on by gravity only is g. Because the force of gravity on the object is the only force acting on it, its weight W (in newtons) is given by:

$$W = mg$$

where m = the mass of the object (in kg).

- When an object is **in equilibrium**, the support force on it is equal and opposite to its weight. Therefore, an object placed on a weighing balance (e.g. a spring balance or a top-pan balance) exerts a force on the balance equal to the weight of the object. Thus the balance measures the weight of the object.

- g is also referred to as the **gravitational field strength** at a given position, as it is the force of gravity per unit mass on a small object at that position. So the gravitational field strength at the Earth's surface is 9.8 N kg^{-1}. Note that the weight of a fixed mass **depends on its location**. For example, the weight of a 1 kg object is 9.8 N on the Earth's surface and 1.6 N on the Moon's surface.

- The mass of an object is a measure of its **inertia**, which is its resistance to change of its motion. Figure 3.6 shows an entertaining demonstration of inertia. When the card is flicked, the coin drops into the glass because the force of friction on it due to the moving card is too small to shift it sideways.

> **The acceleration is always in the same direction as the force.**
>
> *For example, a projectile in motion experiences a force vertically downwards due to gravity. Its acceleration is therefore vertically downwards, no matter what its direction of motion is.*

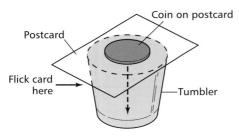

Postcard

Coin on postcard

Flick card here

Tumbler

Fig 3.6 *An 'inertia trick'*

QUESTIONS

1 A car of mass 800 kg accelerates uniformly along a straight line from rest to a speed of 12 m s^{-1} in 50 s. Calculate:

  a the acceleration of the car,

  b the force on the car that produced this acceleration,

  c the ratio of the accelerating force to the weight of the car.

2 An aeroplane, of mass 5000 kg, lands on a runway at a speed of 60 m s^{-1} and stops 25 s later. Calculate:

  a the deceleration of the aeroplane,

  b the braking force on the aeroplane.

3 **a** A vehicle, of mass 1200 kg, on a level road accelerates from rest to a speed of 6.0 m s^{-1} in 20 s, without change of direction. Calculate the force that accelerated the car.

  b The vehicle in part **a** is fitted with a trailer of mass 200 kg. Calculate the time taken to reach a speed of 6.0 m s^{-1} from rest for the same force as in part **a**.

4 A bullet of mass 0.002 kg, travelling at a speed of 120 m s^{-1}, hit a tree and penetrated a distance of 55 mm into the tree. Calculate:

  a the deceleration of the bullet,

  b the impact force of the bullet on the tree.

3.2 Using $F = ma$

3.2.1 Two forces in opposite directions

- When an object is acted on by two **unequal** forces acting in **opposite** directions, the object accelerates in the direction of the larger force. If the forces are F_1 and F_2 where $F_1 > F_2$:

$$\textbf{Resultant force, } \boldsymbol{F_1 - F_2 = ma}$$

where m is the mass of the object and a is its acceleration.

- If the object is on a horizontal surface and F_1 and F_2 are **horizontal** and in opposite directions, the above equation still applies. The support force on the object is equal and opposite to its weight.

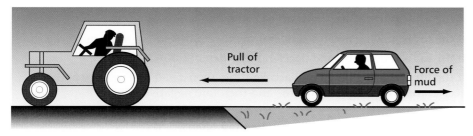

Fig 3.7 *Unbalanced forces*

Some examples are given below, where two forces act in different directions on an object.

Towing a trailer

Consider the example of a car of mass M fitted with a trailer of mass m on a level road. When the car and the trailer accelerate, the car pulls the trailer forward and the trailer holds the car back (Fig 3.8).

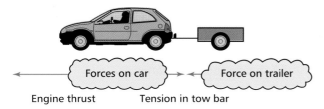

Fig 3.8 *Car and trailer*

- The car is subjected to a driving force F pushing it forwards (from its engine thrust) and the tension T in the tow bar holding it back. Therefore:

$$\textbf{Resultant force on car} = \boldsymbol{F - T = Ma}$$

- The force on the trailer is due to the tension T in the tow bar pulling it forwards. Therefore:

$$T = ma$$

Combining the two equations gives:

$$F = Ma + ma = (M + m)a$$

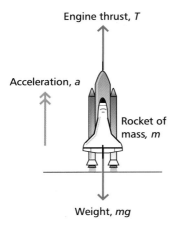

Engine thrust, T

Acceleration, a

Rocket of mass, m

Weight, mg

Fig 3.9 *Rocket launch*

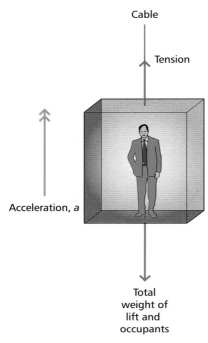

Cable

Tension

Acceleration, a

Total weight of lift and occupants

Fig 3.10 *In a lift*

Rocket problems

If T is the thrust of the rocket engine when its mass is m and the rocket is moving upwards, its acceleration a is given by:

$$T - mg = ma$$

Therefore $$\textbf{\textit{Rocket thrust, T = mg + ma}}$$

The rocket thrust must therefore overcome the weight of the rocket for the rocket to take off (Fig 3.9).

Lift problems

Using 'upwards is positive' gives the resultant force on the lift as:

$$F = T - mg$$

where T is the tension in the lift cable and m is the total mass of the lift and occupants (Fig 3.10).

Therefore $$T - mg = ma$$

where a = acceleration.

- If the lift is moving at a constant velocity, then $a = 0$ so $T = mg$
- If the lift is moving up and accelerating, then $a > 0$ so $T = mg + ma > mg$.
- If the lift is moving up and decelerating, then $a < 0$ so $T = mg + ma < mg$.
- If the lift is moving down and accelerating, then $a < 0$ (i.e. velocity and acceleration are both downwards and therefore negative) so $T = mg - ma < mg$.
- If the lift is moving down and decelerating, then $a > 0$ (i.e. velocity downwards and acceleration upwards and therefore positive) so $T = mg + ma > mg$.

Tension in the cable is less than the weight if:

- The lift is moving up and decelerating (i.e. velocity > 0 and acceleration < 0).
- The lift is moving down and accelerating (i.e. velocity < 0 and acceleration < 0).

Tension in the cable is greater than the weight if:

- The lift is moving up and accelerating (i.e. velocity > 0 and acceleration > 0).
- The lift is moving down and decelerating (i.e. velocity < 0 and acceleration > 0).

Worked example

$g = 9.8 \text{ m s}^{-2}$

A lift of total mass 650 kg moving downwards decelerates at 1.5 m s^{-2} and stops. Calculate the tension in the lift cable during the deceleration.

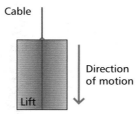

Cable

Direction of motion

Lift

Fig 3.11

Solution

The lift is moving down so its velocity $v < 0$. Since it is decelerating, its acceleration a is in the opposite direction to its velocity, so $a > 0$. Therefore use $a = +1.5 \text{ m s}^{-2}$ in the equation:

$$T - mg = ma$$
$$T = mg + ma$$
$$= 650 \times 9.8 + 650 \times 1.5 = 7300 \text{ N}$$

3.2.2 Further $F = ma$ problems

Pulley problems

Consider two masses M and m (where $M > m$) attached to a thread hung over a frictionless pulley, as in Figure 3.12. When released, mass M accelerates downwards and mass m accelerates upwards. If a is the acceleration and T is the tension in the thread, then:

- On mass M, the resultant force $= Mg - T = Ma$.
- On mass m, the resultant force $= T - mg = ma$.

Therefore, adding the two equations gives:

$$Mg - mg = (M + m)a$$

Sliding down a slope

Consider a block of mass m sliding down a slope (Fig 3.13). The component of the block's weight down the slope is $mg \sin \theta$. If the force of friction on the block is F_0, then:

$$\text{Resultant force on the block} = mg \sin \theta - F_0$$

Therefore $\qquad mg \sin \theta - F_0 = ma$

where a is the acceleration of the block.

Note With the addition of an engine force F_E, the above equation can be applied to a vehicle on a downhill slope of constant gradient. Thus:

$$F_E + mg \sin \theta - F_0 = ma$$

where F_0 is the combined sum of the force of friction and the braking force.

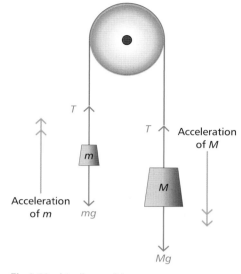

Fig 3.12 A pulley problem

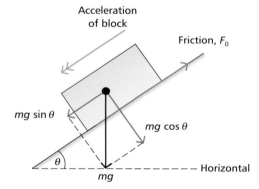

Fig 3.13

QUESTIONS

$g = 9.8 \text{ m s}^{-2}$

1 A rocket of mass 550 kg blasts vertically from the launch pad at an acceleration of 4.2 m s^{-2}. Calculate:

   a the weight of the rocket,

   b the thrust of the rocket engines.

2 A car of mass 1400 kg, pulling a trailer of mass 400 kg, accelerates from rest to a speed of 9 m s^{-1} in a time of 60 s on a level road. Assuming air resistance is negligible, calculate:

   a the tension in the tow bar,

   b the engine force.

3 A lift and its occupants have a total mass of 1200 kg. Calculate the tension in the lift cable when the lift is:

   a stationary,

   b ascending at constant speed,

   c ascending at a constant acceleration of 0.4 m s^{-2},

   d descending at a constant deceleration of 0.4 m s^{-2}.

4 A brick, of mass 3.2 kg on a sloping flat roof, at 30° to the horizontal, slides at constant acceleration 2.0 m down the roof in 2.0 s from rest. Calculate:

   a the acceleration of the brick,

   b the frictional force on the brick due to the roof.

3.3 Terminal speed

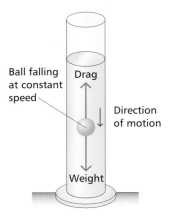

Ball falling at constant speed

Drag

Direction of motion

Weight

a *Falling in a fluid*

b *Skydiving*

Fig 3.14 *At terminal speed*

The acceleration at any instant is the gradient of the speed–time curve.

3.3.1 Drag forces

Any object moving through a fluid experiences a **drag force** due to the fluid. The drag force depends on:

- the shape of the object,
- its speed, and
- the **viscosity** of the fluid (which is a measure of how easily the fluid flows past a surface).

The faster an object travels in a fluid, the greater the drag force on it.

Motion of an object falling in a fluid

- The speed of an object released from rest in a fluid increases as it falls, so the drag force on it due to the fluid increases. The resultant force on the object is the difference between the force of gravity on it (i.e. its weight) and the drag force. As the drag force increases, the resultant force decreases so the acceleration becomes less as it falls. If it continues falling, it attains **terminal speed** when the drag force on it is equal and opposite to its weight. Its acceleration is then zero and its speed remains constant as it falls.

- Figure 3.15 shows how to investigate the motion of an object falling in a fluid. When the object is released, the thread attached to the object pulls a tape through a ticker timer which prints dots on the tape. The spacing between successive dots is a measure of the speed of the object, as the dots are printed on the tape at a constant rate. A tape chart can be made from the tape to show how the speed changes with time. The results show that:

The speed increases and reaches a constant value, which is the terminal speed.

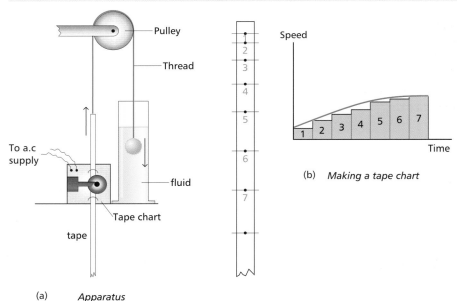

(a)   *Apparatus*

(b)   *Making a tape chart*

Fig 3.15 *Investigating the motion of an object falling in a fluid*

- At any instant, the resultant force is $mg - F_D$, where m is the mass of the object and F_D is the drag force.

Therefore   **Acceleration of the object** $= \dfrac{mg - F_D}{m} = \dfrac{g - F_D}{m}$

1 The **initial acceleration** is g, because the drag force is zero at the instant the object is released.

2 At the **terminal speed**, the potential energy lost by the object as it falls is converted to internal energy of the fluid by the drag force.

Motion of a powered vehicle

The top speed of a road vehicle or an aircraft depends on its engine power, also referred to as its **motive power**, and its shape. A vehicle with a streamlined shape can reach a higher top speed than a vehicle with the same motive power that is not streamlined.

For a powered vehicle of mass m moving on a level surface, if F_E represents the motive force of the vehicle (i.e. the engine force driving the vehicle):

$$\textbf{Resultant force} = F_E - F_D$$

where F_D is the drag force.

Therefore, its acceleration $\qquad a = \dfrac{(F_E - F_D)}{m}$

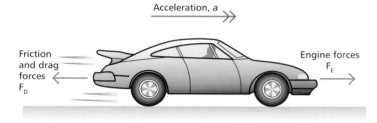

Fig 3.16

- Because the drag force increases with speed, the maximum speed (i.e. terminal speed) of the vehicle, v_{max}, is attained when the drag force becomes equal and opposite to the motive force.

- At maximum speed, the work done by the engine is dissipated by the drag force and becomes internal energy of the surroundings. See section 4.3 for more about motive power.

Worked example

A car of mass 1200 kg has an engine which provides a motive force of 600 N. Calculate:

a its initial acceleration,

b its acceleration when the drag force is 400 N.

Solution

a The maximum acceleration is when the drag force is zero (i.e. when the car starts). The resultant force on the car is therefore 600 N at the start.

Therefore

$$\text{initial acceleration} = \frac{\text{force}}{\text{mass}} = \frac{600 \text{ N}}{1200 \text{ kg}} = 0.5 \text{ m s}^{-2}$$

b When the drag force = 400 N,
the resultant force $\quad$ = weight − drag force
$\qquad\qquad\qquad\qquad$ = 600 − 400
$\qquad\qquad\qquad\qquad$ = 200 N

∴ Acceleration = force/mass = $\dfrac{200 \text{ N}}{1200 \text{ kg}}$ = 0.16 m s^{-2}

Hydrofoil physics

A hydrofoil boat travels much faster than an ordinary boat, because it has a powerful jet engine that enables it to 'ski' on its hydrofoils when the jet engine is switched on.

When the jet engine is switched on and takes over from the less-powerful propeller engine, the boat speeds up and the hydrofoils are extended. The boat rides on the hydrofoils so the drag force is reduced, as its hull is no longer in the water. At top speed, the motive force of the jet engine is equal to the drag force on the hydrofoils.

Fig 3.17 *A hydrofoil ferry*

QUESTIONS

$g = 9.8$ m s^{-2}.

1 a A steel ball of mass 0.15 kg released from rest in a liquid falls a distance of 0.20 m in 5.0 s. Assuming the ball reaches terminal speed within a fraction of a second, calculate:
  (i) its terminal speed,
  (ii) the drag force on it when it falls at terminal speed.

  b State and explain whether or not a smaller steel ball would fall at the same rate in the same liquid.

2 Explain why a cyclist can reach a higher top speed by crouching over the handlebars instead of sitting upright while pedalling.

3 A vehicle of mass 32 000 kg has an engine which has a maximum driving force of 4.4 kN and a top speed of 36 m s^{-1} on a level road.
Calculate:

  a its maximum acceleration from rest,

  b the distance it would travel at maximum acceleration to reach a speed of 12 m s^{-1} from rest.

4 Explain why a vehicle has a higher top speed on a downhill stretch of road than on a level road.

3.4 On the road

3.4.1 Stopping distances

Traffic accidents often happen because vehicles are being driven too fast and too close to the vehicle in front. A driver needs to maintain a safe distance between his or her own vehicle and the vehicle in front. If a vehicle suddenly brakes, the driver of the following vehicle needs to brake as well to avoid a crash.

- **Thinking distance** is the distance travelled by a vehicle in the time it takes the driver to react. For a vehicle moving at constant speed v:

$$\textbf{Thinking distance, } s_1 = \textbf{speed} \times \textbf{reaction time}$$
$$= vt_0$$

where t_0 is the reaction time of the driver.

- **Braking distance** is the distance travelled by a car in the time it takes to stop safely from when the brakes are first applied. Assuming constant deceleration, a, to zero speed from speed v:

$$\textbf{Braking distance, } s_2 = \frac{v^2}{2a}$$

since $v^2 = 2as_2$

- **Stopping distance** is the sum of the braking distance and the thinking distance:

$$\textbf{Stopping distance} = vt_0 + \frac{v^2}{2a}$$

where v is the speed before the brakes were applied.

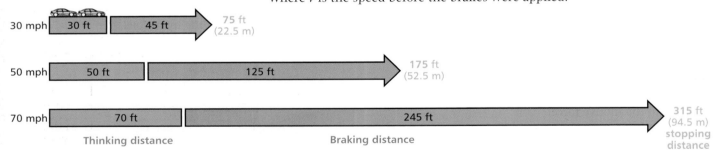

30 mph | 30 ft | 45 ft | 75 ft (22.5 m)

50 mph | 50 ft | 125 ft | 175 ft (52.5 m)

70 mph | 70 ft | 245 ft | 315 ft (94.5 m) stopping distance

Thinking distance   Braking distance

Fig 3.18 Stopping distances

Figure 3.18 shows how thinking distance, braking distance and stopping distance vary with speed for a reaction time of 0.67 s and a deceleration of 6.75 m s^{-2}. Using these values for reaction time and deceleration in the above equations gives the shortest stopping distances on a dry road as recommended in the Highway Code.

3.4.2 Skidding

- On a front-wheel drive vehicle, the front wheels are driven by the engine via the transmission system. The driving wheels generate a **motive force** which drives the vehicle forward. This is because the engine turns the driving wheels and friction between the tyres and the road prevent 'wheel spin' (i.e. slipping), so the driving wheels 'roll' along the road. If the driver tries to accelerate too fast, the wheels skid. This is because there is an upper limit to the amount of friction between the tyres and the road.

- When the brakes are applied, the wheels are slowed down by the brakes. The vehicle therefore slows down, provided the wheels do not skid. If the braking force is increased, the friction force between the tyres and the road increases.

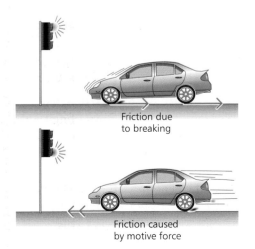

Friction due to breaking

Friction caused by motive force

Fig 3.19 Stopping and starting

However, if the upper limit of friction (usually referred to as 'limiting frictional force') between the tyres and the road is reached, the wheels skid. When this happens, the brakes lock and the vehicle slides uncontrollably forward.

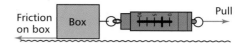

Fig 3.20 *Testing friction*

Testing friction

Figure 3.20 shows how to measure the limiting friction between two horizontal surfaces. If the block is pulled with an increasing force, the block does not move until the force pulling it is enough to overcome the limiting frictional force between it and the surface supporting it. Before this force is reached, the block is prevented from moving by the frictional force between it and the surface. The **limiting frictional force** on the block is therefore equal to the **pull force** on the block just before sliding occurs.

More about braking distance

The braking distance for a vehicle depends on the **speed** of the vehicle at the instant the brakes are applied, on the **road conditions** and on the **condition of the vehicle tyres**.

- **On a greasy or icy road**, skidding is more likely because the limiting frictional force is reduced from its 'dry' value. Therefore, to stop a vehicle safely on a greasy or icy road, the brakes must be applied with less force than on a 'dry' road – otherwise skidding will occur. Therefore the **braking distance** (i.e. the distance to stop safely) is longer than on a dry road. In fast-moving traffic, a driver must ensure there is a bigger gap to the car in front so he or she can slow down safely if the vehicle in front slows down.

- The condition of the tyres of a vehicle also affects braking distance. The **tread of a tyre** must not be less than a certain depth otherwise any grease, oil or water on the road reduces friction very considerably. In the United Kingdom, the absolute mimimum legal requirement is that every tyre of a vehicle must have a tread depth of at least 1 mm across three-quarters of the width of the tyre all the way round. If the pressure on a tyre is too small, or too great, or the wheel is 'unbalanced', then the tyre will wear unevenly and will quickly become unsafe. Clearly, a driver must check the condition of the tyres regularly. If you are a driver, or hope to become one, remember that the vehicle tyres are the only contacts between the vehicle and the road and must therefore be looked after.

Fig 3.21 *Tyre treads*

Worked example

$g = 9.8 \text{ m s}^{-2}$

A vehicle of mass 900 kg on a level road, travelling at a speed of 15 m s^{-1} can be brought to a standstill, without skidding, by a braking force equal to 0.5 × its weight. Calculate:

a the deceleration of the vehicle,

b the braking distance.

Solution

a Weight = 900 × 9.8 = 8800 N

Braking force = 0.5 × 8800 = 4400 N

$$\text{Deceleration} = \frac{\text{braking force}}{\text{mass}} = \frac{4400}{900} = 4.9 \text{ m s}^{-2}$$

b To calculate *s*, use $v^2 = u^2 + 2as$

$u = 15 \text{ m s}^{-1}, v = 0, a = -4.9 \text{ m s}^{-2}$

$$\therefore s = -\frac{u^2}{2a} = \frac{15^2}{2 \times 4.9} = 23 \text{ m}$$

QUESTIONS

1 A vehicle is travelling at a speed of 18 m s^{-1} on a level road when the driver sees a pedestrian stepping off the pavement into the road, 45 m ahead. The driver reacts within 0.4 s and applies the brakes, causing the car to decelerate at 4.8 m s^{-2}.

  a Calculate: (i) the thinking distance,
                   (ii) the braking distance.

  b How far does the driver stop from where the pedestrian stepped into the road ?

2 The braking distance of a vehicle for a speed of 18 m s^{-1} on a dry level road is 24 m. Calculate:

  a the deceleration of the vehicle from this speed to a standstill over this distance,

  b the frictional force on a vehicle of mass 1000 kg on this road when it stops.

3 a What is meant by the braking distance of a vehicle ?

  b Explain, in terms of the forces acting on the wheels of a car, why a vehicle slows down when the brakes are applied.

4 The frictional force on a vehicle travelling on a certain type of road surface is 0.6 × the vehicle's weight when the road is level. For a vehicle of mass 1200 kg:

  a Show that the maximum decleration on this road is 5.9 m s^{-2}.

  b Calculate the braking distance on this road for a speed of 30 m s^{-1}.

Vehicle safety

Fig 3.22 *A car crash*

Fig 3.23 *A video clip of a side-on impact*

3.5.1 Impact forces

Measuring impacts

The **effect of a collision** on a vehicle can be measured in terms of the acceleration or deceleration of the vehicle. By expressing an acceleration or deceleration in terms of g, the acceleration due to gravity, the force of the impact can then easily be related to the weight of the vehicle. For example, suppose a vehicle hits a wall and its deceleration is -30 m s^{-2}. In terms of g, the deceleration is $3g$. So the force of the wall on the vehicle must have been three times its weight ($= 3\,mg$, where m is the mass of the vehicle). Such an impact is sometimes described as being 'equal to $3g$'. This statement, although technically wrong because the acceleration not the impact is equal to $3g$, is a convenient way of expressing the effect of an impact on a vehicle or a person.

How much acceleration or deceleration can a person withstand? The **duration** of the impact, as well as the **magnitude** of the acceleration, affects the person. A person who is sitting or upright can survive a deceleration of $20g$ for a time of a few milliseconds, although not without severe injury. A deceleration of over $5g$ lasting for a few seconds can cause injuries. Car designers carry out tests using dummies in remote-control vehicles, to measure the change of motion of different parts of a vehicle or a dummy. Sensors linked to data recorders and computers are used, as well as video cameras, to record the motion for video clips to be analysed.

Contact time and impact distance

When objects collide and bounce off each other, they are in contact with each other for a certain time which is the same for both objects. The shorter the **contact time**, the greater the impact force for the same initial velocities of the two objects. When two vehicles collide, they may or may not separate from each other after the collision. If they remain tangled together, they exert forces on each other until they are moving at the same velocity. The duration of the impact force is not the same as the contact time in this situation.

The impact time, t, (i.e. the duration of the impact force) can be worked out using the 'suvat' equation $s = \frac{1}{2}(u + v)\,t$ applied to one of the vehicles, where s is the distance moved by that vehicle during the impact, u is its initial velocity and v is its final velocity.

For example, suppose a vehicle moving at 20 m s^{-1} slows down in a distance of 4.0 m to a velocity of 12 m s^{-1} as a result of hitting a stationary vehicle. Rearranging the above equation gives:

$$t = \frac{2s}{(u + v)} = \frac{2 \times 4.0}{(20 + 12)} = 0.25 \text{ s}$$

The deceleration, $a = \dfrac{(v - u)}{t} = \dfrac{(12 - 20)}{0.25} = -32$ m s^{-2}

$$= 3.3g \text{ (where } g = 9.8 \text{ m s}^{-2}\text{)}$$

Note The work done by the **impact force**, F, over an impact distance s ($= Fs$) is equal to the change of kinetic energy of the vehicle. The impact force can also be worked out using the equation:

$$F = \frac{\text{change of kinetic energy}}{\text{impact distance}}$$

3.5.2 Car safety features

In the above example the vehicle's deceleration, and hence the impact force, would be lessened if the **impact time** were greater. So by increasing the impact time, the impact force is reduced. The following vehicles safety features are designed to increase the impact time and so reduce the impact force. With a reduced impact force, the vehicle occupants are less affected.

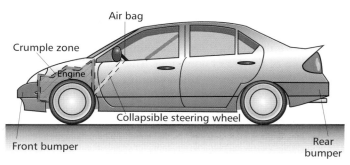

Fig 3.24 *Vehicle safety features*

- **Vehicle 'bumpers'** give way a little in a low-speed impact and so increase the impact time. The impact force is therefore reduced as a result. If the initial speed of impact is too high, the bumper and/or the vehicle chassis is likely to be damaged.

- **Crumple zones** are where the engine compartment of a car is designed to give way in a front-end impact. If the engine compartment were rigid, the impact time would be very short and the impact force would be very large. By designing the engine compartment so it 'crumples' in a front-end impact, the impact time is increased and the impact force is therefore reduced.

- In a front-end impact, a correctly-fitted **seat belt** restrains the wearer from crashing into the vehicle frame after the vehicle suddenly stops. The restraining force on the wearer is therefore much less than the impact force would be if the wearer hit the vehicle frame. With the seat belt on, the wearer is stopped more gradually than without it.

- **Collapsible steering wheels** are helpful in front-end impacts, where the seat belt restrains the driver without holding him or her rigidly. If the driver makes contact with the steering wheel, the impact force is lessened as a result of the steering wheel collapsing in the impact.

- An **air bag** reduces the force on a person, because the air bag acts as a cushion and increases the impact time on the person. More significantly, the force of the impact is spread over the contact area, which is greater than the contact area with a seat belt. So the pressure on the body is less.

Fig 3.25 *An air bag in action*

QUESTIONS

$g = 9.8 \text{ m s}^{-2}$

1 A car of mass 1200 kg travelling at a speed of 15 m s^{-1} is struck from behind by another vehicle, causing its speed to increase to 19 m s^{-1} in a time of 0.20 s. Calculate:
   a the acceleration of the car, in terms of g,
   b the impact force on the car.

2 The front end of a certain type of car of mass 1500 kg, travelling at a speed of 20 m s^{-1}, is designed to crumple in a distance of 0.8 m if the car hits a wall. Calculate:
   a the impact time if it hits a wall at 20 m s^{-1},
   b the impact force.

3 The front bumper of a car of mass 900 kg is capable of withstanding an impact with a stationary object, provided the car is not moving faster than 3.0 m s^{-1} when the impact occurs. The impact time at this speed is 0.40 s. Calculate:
   a the deceleration of the car from 3.0 m s^{-1} to rest in 0.40 s,
   b the impact force on the car.

4 In a crash, a vehicle travelling at a speed of 25 m s^{-1} is stopped in a distance of 4.5 m. A passenger of mass 68 kg travelling in the vehicle is wearing a seat belt, which restrains her forward movement relative to the car to a distance of 0.5 m.
   a Show that the deceleration of the passenger is 62.5 m s^{-2}.
   b Calculate the deceleration of the passenger in terms of g.

1 a Explain why a force is needed to make an object move at a steady speed across a rough floor.

   b Explain why an object sliding across ice keeps moving even though no force is pushing it.

2 A vehicle of mass 450 kg accelerates from rest to a velocity of 9.5 m s^{-1} in 38 s.

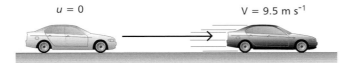

$u = 0$ $V = 9.5$ m s^{-1}

Calculate:

a the acceleration of the vehicle,

b the motive force of the vehicle engine,

c the ratio of the motive force to the vehicle's weight.

3 A bullet of mass 2.4×10^{-3} kg moving at a speed of 115 m s^{-1} hits a tree and embeds itself in the tree to a depth of 85 mm.

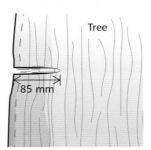

Tree

85 mm

Calculate:

a the impact time,

b the deceleration of the bullet in the tree,

c the impact force on the bullet.

4 A heavy goods vehicle of total weight 320 kN has an engine which has a maximum motive force of 6 kN.

   a Calculate:
   (i) the mass of the vehicle,
   (ii) the maximum acceleration of the vehicle,
   (iii) the time taken for the vehicle to accelerate from rest to a speed of 20 m s^{-1},
   (iv) the distance moved by the vehicle in this time.

   b The vehicle is used to pull a trailer of weight 60 kN. Calculate:
   (i) the maximum acceleration of the vehicle when the trailer is attached,
   (ii) the time taken by the vehicle to reach a speed of 20 m s^{-1} from rest with the trailer attached.

5 A steel ball bearing was released in water and fell to the bottom of the water container. The time of descent was measured at different distances as it fell. The figure below shows the measurements plotted as a graph of distance fallen against time taken.

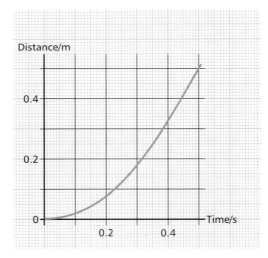

   a (i) What feature of this graph represents the speed of the ball bearing at any instant?
   (ii) Describe how the speed of the ball bearing changed as it fell.
   (iii) Sketch a graph to show how the speed of the ball bearing changed with time during its descent.

   b Explain, in terms of the forces acting on the ball, why the acceleration of the ball bearing decreased as it fell.

6 a (i) What is meant by the 'thinking distance' of a driver ?

    (ii) The thinking distance of the driver in a car travelling at 18 m s⁻¹ is 12 m. Calculate the reaction time of the driver.

b (i) What is meant by the 'braking distance' of a car?

    (ii) The driver of a car travelling at a speed of 18 m s⁻¹ applies the brakes and stops in a braking distance of 24 m. Calculate the deceleration of the car.

    (iii) If the car in part (ii) has a total mass of 750 kg, calculate the braking force needed to stop the car.

7 a Explain why each of the following vehicle safety features reduces the risk of injury in a car accident:
    (i) seat belts,   (ii) crumple zones.

b In a test using a dummy of mass 64 kg in a vehicle designed with crumple zones, the vehicle was driven straight into a large concrete wall to observe the effect on the dummy. The dummy was fitted with a seat belt. When the vehicle was driven at the wall at a speed of 8 m s⁻¹, the dummy was stopped in a time of 0.4 s. Calculate:
    (i) the distance moved by the dummy in this time,
    (ii) the deceleration of the dummy,
    (iii) the force of the seat belt on the dummy.

c Discuss what would have been the effect on the dummy if the crash had taken place without the dummy having been fitted with a seat belt.

8 A parachutist of total mass 65 kg is in free fall at constant speed when she opens her parachute.

a (i) Calculate the weight of the parachutist.
    (ii) Calculate the drag force on the parachutist before she opened her parachute.

b The figure below shows how her speed changed after her parachute opened.

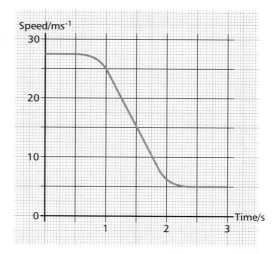

    (i) Explain why her speed decreased suddenly when she opened her parachute.
    (ii) Use the graph to estimate her maximum deceleration after she opened the parachute.
    (iii) Sketch a graph to show how the drag force on the parachutist changed as a result of opening the parachute.

4 *Work, energy and power*

4.1

Work and energy

4.1.1 Energy rules

Energy is needed to make stationary objects move, or to lift an object, or to change its shape, or to warm it up. When you lift an object, you **transfer energy** from your muscles to the object.

Objects can possess energy in **different forms**, including:
- **potential energy**, which is energy due to position;
- **kinetic energy**, which is energy due to motion;
- **thermal energy**, which is energy due to the temperature of an object;
- **chemical** or **nuclear energy**, which is energy associated with chemical or nuclear reactions;
- **electrical energy**, which is energy of electrically charged objects;
- **elastic energy**, which is energy stored in an object when it is stretched or compressed.

Energy is measured in **joules (J)**. One joule is equal to the energy needed to raise a 1 N weight through a vertical height of 1 m.

Energy can be changed from one form into other forms. In any change, the total amount of energy after the change is always equal to the total amount of energy before the change. The total amount of energy is unchanged. In other words:

Energy cannot be created or destroyed.

This statement is known as the **Principle of Conservation of Energy**.

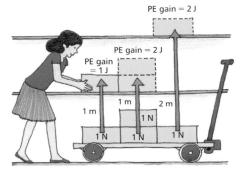

Fig 4.1 Using joules

Work
Work is done on an object when a force acting on it due to another object makes it move. As a result, energy is transferred to the object by the force from the other object. The amount of work done depends on the force and the distance moved by the object. The greater the force or the further the distance, the greater the work done.

Work done = force × distance moved in the direction of the force

The **unit of work** is the joule (J).

One joule is equal to the work done when a force of 1 N moves its point of application by a distance of 1 m in the direction of the force.

Energy is also measured in joules, because the energy transferred to an object when a force acts on it is equal to the work done on it by the force. For example, a force of 2 N needs to be applied to an object of weight 2 N to raise the object steadily. If the object is raised by 1.5 m, the work done by the force is 3 J (= 2 N × 1.5 m). Therefore, the gain of potential energy of the object is 3 J.

Force and displacement
Imagine a yacht acted on by a wind force F at an angle θ to the direction in which the yacht moves. The wind force has a component $F\cos\theta$ in the direction of motion of the yacht, and a component $F\sin\theta$ at right angles to the direction of motion. If the yacht is moved a distance s by the wind, the work done on it is equal to the component of force in the direction of motion × the distance moved:

$$W = Fs \cos \theta$$

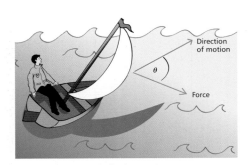

Fig 4.2 Force and displacement

Note *If $\theta = 90°$, which means that the force is perpendicular to the direction of motion, then, because $\cos 90° = 0$, the work done is zero.*

4.1.2 Force-distance graphs

- If a constant force F acts on an object and makes it move a distance s in the direction of the force, the work done on the object $W = Fs$. Figure 4.3 shows a graph of force against distance in this situation. The area under the line is a rectangle, of height representing the force and of base length representing the distance moved. Therefore **the area represents the work done**.

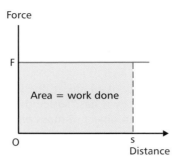

Fig 4.3 *A force-distance graph for a constant force*

- If a variable force F acts on an object and causes it to move in the direction of the force, for a small amount of distance Δs, the work done $\Delta W = F\,\Delta s$. This is represented on a graph of the force F against distance s by the area of a strip under the line of width Δs and height F. The total work done is therefore **the sum of the areas of all the strips** (i.e. the total area under the line).

The area under the line of a force-distance graph represents the total work done.

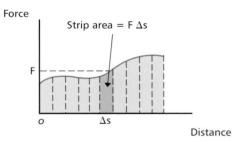

Fig 4.4 *A force-distance graph for a variable force*

Springs

For example, consider the force needed to stretch a spring. The greater the force, the more the spring is extended from its unstretched length. Figure 4.5 shows how the force needed to stretch a spring changes with the **extension** of the spring. The graph is a straight line through the origin. Therefore the force needed is proportional to the extension of the spring. This is known as **Hooke's Law**. See section 5.3 for more about springs.

Figure 4.5 is a graph of force against distance, in this case the distance the spring is extended. As explained above, the area under the line represents the work done to stretch the spring. Let F_0 represent the force needed to extend the spring to extension e_0. Therefore, because the area under the line from the origin to extension e_0 is a triangle of height F_0 and base length e_0:

$$\textbf{Area under line} = \tfrac{1}{2} \times \textbf{height} \times \textbf{base} = \tfrac{1}{2}F_0 e_0$$

$$\textbf{Work done to stretch spring to extension } e_0 = \tfrac{1}{2}F_0 e_0$$

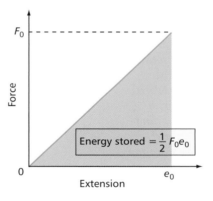

Fig 4.5 *Force against extension for a spring*

QUESTIONS

1 Calculate the work done when:

  a a weight of 40 N is raised by a height of 5.0 m,

  b a spring is stretched to an extension of 0.45 m by a force of 20 N.

2 Calculate the energy transferred by a force of 12 N when it moves an object by a distance of 4.0 m:

  a in the direction of the force,

  b in a direction at 60° to the direction of the force,

  c in a direction at right angles to the direction of the force.

3 A luggage trolley of total weight 400 N is pushed at a steady speed 20 m up a slope by a force of 50 N acting in the same direction as the object moves in. At the end of this distance, the trolley is 1.5 m higher than at the start. Calculate:

  a the work done pushing the trolley up the slope,

  b the gain of potential energy of the trolley,

  c the energy wasted due to friction.

4 A spring that obeys Hooke's Law requires a force of 1.2 N to extend it to an extension of 50 mm. Calculate:

  a the force needed to extend it to an extension of 100 mm,

  b the work done when the spring is stretched to an extension of 100 mm.

4.2   Kinetic energy and potential energy

4.2.1   Kinetic energy

Kinetic energy is the energy of an object due to its motion. The faster an object moves, the more kinetic energy it has. To see the exact link between kinetic energy and speed, consider an object of mass m, initially at rest, acted on by a constant force F for a time t.

Fig 4.6 *Gaining kinetic energy*

Let the speed of the object at time t be v:

$$\therefore \qquad \text{Distance travelled, } s = \tfrac{1}{2}(u + v)t$$
$$= \tfrac{1}{2}vt \text{ because } u = 0$$
$$\text{Acceleration, } a = \frac{(v - u)}{t} = \frac{v}{t}$$

Using Newton's Second Law:

$$F = ma = \frac{mv}{t}$$

$\therefore$ the work done, by force F, to move the object through distance s:

$$W = Fs = \frac{mv}{t} \times \frac{vt}{2} = \tfrac{1}{2}mv^2$$

Because the gain of kinetic energy is due to the work done, then

$$\textbf{Kinetic energy, } E_K = \tfrac{1}{2}mv^2$$

Note *The formula does not hold at speeds* **approaching the speed of light**. *Einstein's theory of special relativity tells us that the mass of an object increases with speed and that the energy of an object can be worked out from the equation* $E = mc^2$, *where c is the speed of light in free space and m is the mass of the object.*

4.2.2   Potential energy

Potential energy is the energy of an object due to its position.
If an object of mass m is raised through a vertical height h at steady speed, the force needed to raise it is equal and opposite to its weight mg. Therefore:

Work done to raise the object = force × distance moved = mgh

The work done on the object increases its gravitational potential energy:

Change of gravitational potential energy, $E_P = mgh$

At the Earth's surface, $g = 9.8 \text{ m s}^{-2}$.

Note *The formula does not hold unless the change of height h is much smaller than the* **Earth's radius**. *If height h is not insignificant compared with the Earth's radius, the value of g is not the same over height h. The force of gravity on an object decreases with increased distance from the Earth.*

4.2.3   Energy changes involving kinetic and potential energy

An object of mass m released above the ground
If air resistance is negligible, the object gains speed as it falls. Its potential energy therefore decreases and its kinetic energy increases.

After falling through a vertical height h, its kinetic energy is equal to its loss of potential energy.

In other words: $\qquad\qquad\qquad \tfrac{1}{2}mv^2 = mgh$

A pendulum bob

A pendulum bob is displaced from equilibrium and then released with the thread taut. The bob passes through equilibrium at maximum speed, and then slows down to reach maximum height on the other side of equilibrium. If its initial height above equilibrium is h_0, then whenever its height above equilibrium is h, its speed v at this height is such that:

Its kinetic energy = its loss of potential energy from maximum height

$$\tfrac{1}{2} mv^2 = mgh_0 - mgh$$

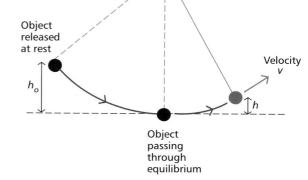

Velocity
v

Object released at rest

h_o

Object passing through equilibrium

Fig 4.7 *A pendulum in motion*

A fairground vehicle of mass *m* on a downward track

If a fairground vehicle was initially at rest at the top of the track, and its speed is v at the bottom of the track, then at the bottom of the track:

- its kinetic energy $= \tfrac{1}{2} mv^2$,
- its loss of potential energy $= mgh$, where h is the vertical distance between the top and the bottom of the track,
- the work done to overcome friction and air resistance $= mgh - \tfrac{1}{2} mv^2$.

Worked example

$g = 9.8 \text{ m s}^{-2}$

On a fairground ride, the track descends by a vertical drop of 55 m over a distance of 120 m along the track. A train of mass 2500 kg on the track reaches a speed of 30 m s^{-1} at the bottom of the descent, after being at rest at the top. Calculate:

a the loss of potential energy of the train,
b its gain of kinetic energy,
c the average frictional force during the descent.

Solution

a Loss of potential energy $= mgh = 2500 \times 9.8 \times 55 = 1.35 \times 10^6$ J

b Its gain of kinetic energy $= \tfrac{1}{2} mv^2 = 0.5 \times 2500 \times 30^2 = 1.13 \times 10^6$ J

c Work done to overcome friction $= mgh - \tfrac{1}{2} mv^2$
$$= 1.35 \times 10^6 - 1.13 \times 10^6$$
$$= 2.4 \times 10^5 \text{ J}$$

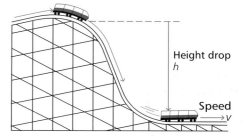

Height drop
h

Speed
v

Fig 4.8

Work done to overcome friction = frictional force × distance moved along track

So Frictional force $= \dfrac{\text{work done to overcome friction}}{\text{distance moved}} = \dfrac{2.4 \times 10^5 \text{ J}}{120 \text{ m}} = 2000$ N

QUESTIONS

1 A ball of mass 0.50 kg was thrown directly up at a speed of 6.0 m s^{-1}. Calculate:
  a its kinetic energy at 6 m s^{-1},
  b its maximum gain of potential energy,
  c its maximum height gain.

2 A ball of mass 0.20 kg, at a height of 1.5 m above a table, is released from rest and it rebounds to a height of 1.2 m above the table. Calculate:
  a (i) the loss of potential energy on descent,
  (ii) the gain of potential energy at maximum rebound height,
  b the loss of kinetic energy due to the impact.

3 A cyclist of mass 80 kg (including the bicycle) freewheels from rest 500 m down a hill. The foot of the hill is 20 m lower than the cyclist's starting point, and the cyclist reaches a speed of 12 m s^{-1} at the foot of the hill. Calculate:
  a (i) the loss of potential energy,
  (ii) the gain of kinetic energy of the cyclist and cycle,
  b (i) the work done against friction and air resistance during the descent,
  (ii) the average resistive force during the descent.

4 A fairground vehicle of total mass 1200 kg, moving at a speed of 2 m s^{-1}, descends through a height of 50 m to reach a speed of 28 m s^{-1} after travelling a distance of 75 m along the track. Calculate:
  a its loss of potential energy,
  b its initial kinetic energy,
  c its kinetic energy after the descent,
  d the work done against friction,
  e the average frictional force on it during the descent.

Power

4.3.1 Power and energy

Energy transfers

Energy can be **transferred** from one object to another by means of:

- **Work done** by a force due to one object making the other object move.
- **Heat transfer** from a hot object to a cold object. Heat transfer can be due to conduction, convection or radiation.
- In addition, **electricity**, **sound waves and electromagnetic radiation**, such as light or radio waves, transfer energy.

In any energy transfer process, the more energy transferred per second, the greater the **power** of the transfer process. For example, in a tall building where there are two elevators of the same total weight, the more powerful elevator is the one that can reach the top floor faster. In other words, its motor transfers energy from electricity at a faster rate than the motor of the other elevator. The energy transferred per second is the **power** of the motor:

> **Power is defined as the rate of transfer of energy.**

The unit of power is the watt (W), equal to an energy transfer rate of 1 joule per second.

Note 1 kilowatt (kW) = 1000 W, and 1 megawatt (MW) = 10^6 W

If energy E is transferred steadily in time t:

$$\textbf{Power, } P = \frac{E}{t}$$

Where energy is transferred by a force doing work, the energy transferred is equal to the work done by the force. Therefore, the **rate of transfer of energy** is equal to the work done per second. In other words, if the force does work W in time t:

$$\textbf{Power, } P = \frac{W}{t}$$

Power measurements

1 Muscle power

Test your own muscle power by timing how long it takes you to walk up a flight of steps. To calculate your muscle power, you will need to know your weight and the total height gain of the flight of steps:

- Your gain of potential energy = your weight × total height gain

- Your muscle power, $P = \dfrac{\text{energy transferred}}{\text{time taken}} = \dfrac{\text{weight} \times \text{height gain}}{\text{time taken}}$

Worked example

A person of weight 480 N, who climbs a flight of stairs of height 10 m in 12 s, has leg muscles of power: 480 N × 10 m/12 s = 400 W

Each leg would therefore have muscles of power 200 W.

2 Electrical power

The power of a 12 V light bulb can be measured using a joulemeter, as shown in Figure 4.10. The joulemeter is read before and after the light bulb is switched on. The difference between the readings is the energy supplied to the light bulb. If the light bulb is switched on for a measured time, the power of the light bulb can be calculated from the energy supplied to it/the time taken.

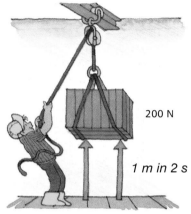

200 N

1 m in 2 s

Fig 4.9 *A 100 watt worker*

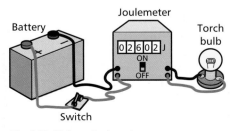

Fig 4.10 *Using a joulemeter*

4.3.2 Motive power

Vehicle engines, marine engines and aircraft engines are all designed to make objects move. The output power of an engine is called its **motive** power.

> When a powered object moves at a constant velocity at a constant height, the resistive forces (e.g. friction, air resistance, drag) are equal and opposite to the motive force.

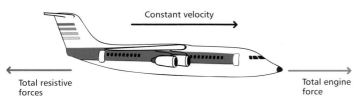

Constant velocity

Total resistive forces

Total engine force

Fig 4.11 Engine power

The work done by the engine is converted to internal energy of the surroundings by the resistive forces.
For a powered vehicle driven by a constant force F moving at speed v:

Work done per second = force × distance moved per second

∴ Motive power of the engine, $P = F v$

Worked example

An aircraft powered by engines that exert a force of 40 kN is in level flight at a constant velocity of 80 m s^{-1}. Calculate the motive power of the engine at this speed.

Solution

Power = force × velocity = 40 000 N × 80 m s^{-1} = 3.2×10^6 W

> When a powered object gains speed, the motive force exceeds the resistive forces on it.

Consider a vehicle that speeds up on a level road. The motive power of its engine is the work done by the engine per second. The work done by the engine increases the kinetic energy of the vehicle and enables the vehicle to overcome the resistive forces acting on. Because the resistive forces increase the internal energy of the surroundings:

Motive power = energy per second wasted + the gain of kinetic energy per second

Juggernaut physics

The maximum weight of a truck on UK roads must not exceed 38 tonnes, which corresponds to a total mass of 38 000 kg. This limit is set so as to prevent damage to roads and bridges. European Union regulations limit the motive power of a large truck to a maximum of 6 kW per tonne. Therefore, the maximum motive power of a 38 tonne truck is 228 kW. Prove for yourself that a truck with a motive power of 228 kW moving at a constant speed of 31 m s^{-1} (= 70 miles per hour) along a level road experiences a drag force of 7.4 kN.

Fig 4.12 Heavy goods on the move

QUESTIONS

$g = 9.8$ m s^{-2}

1 A student of weight 450 N climbed 2.5 m up a rope in 18 s. Calculate:

   a the gain of potential energy of the student,

   b the energy transferred per second.

2 Calculate the power of the engines of an aircraft at a speed of 250 m s^{-1}, if the total engine thrust to maintain this speed is 2.0 MN.

3 A rocket of mass 5800 kg accelerates vertically from rest to a speed of 220 m s^{-1} in 25 s. Calculate:

   a its gain of potential energy,

   b its gain of kinetic energy,

   c the power output of its engine, assuming no energy is wasted due to air resistance.

4 Calculate the height through which a 5 kg mass would need to be dropped to lose the same energy as a 100 W light bulb would use in 1 min.

4.4

More about energy

4.4.1 Machine power

A machine that lifts or moves an object applies a force to the object to move it. If the machine exerts a force F on an object to make it move through a distance s in the direction of the force, the work done W on the object by the machine can be calculated using the equation

Work done, $W = Fs$

If the object moves at a constant velocity v due to this force being opposed by an equal and opposite force caused by friction, the object moves a distance $s = vt$ in time t.

Therefore, the output power of the machine

$$P_{OUT} = \frac{\text{work done by the machine}}{\text{time taken}} = \frac{Fvt}{t} = Fv$$

Output power, $P_{OUT} = Fv$, where $F = $ 'output' force of the machine and $v = $ speed of the object

Examples

1 An electric motor operating a sliding door exerts a force of 125 N on the door, causing it to open at a constant speed of 0.40 m s⁻¹. The output power is 125 N × 0.40 m s⁻¹ = 50 W. The motor must therefore transfer 50 J every second to the sliding door while the door is being opened.

 Friction in the motor bearings and also electrical resistance of the motor wires means that some of the electrical energy supplied to the motor is wasted. For example, if the motor is supplied with electrical energy at a rate of 150 J s⁻¹ and it transfers 50 J s⁻¹ to the door, the difference of 100 J s⁻¹ is wasted as a result of friction and electrical resistance in the motor.

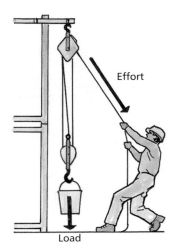

Fig 4.13 Using pulleys

2 A pulley system is used to raise a load of 80 N at a speed of 0.15 m s⁻¹ by means of a constant 'effort' of 30 N applied to the system. Figure 4.13 shows the arrangement. Note that for every metre the load rises, the effort needs to act over a distance of three metres because the load is supported by 3 sections of rope. The effort must therefore act at a speed of 0.45 m s⁻¹ (= 3 × 0.15 m s⁻¹).

 • The work done on the load each second = load × distance per second
  = 80 N × 0.15 m s⁻¹ = 12 J s⁻¹
 • The work done by the effort each second = effort × distance each second
  = 30 N × 0.45 m s⁻¹ = 13.5 J s⁻¹

 The difference of 1.5 J s⁻¹ is the energy wasted each second in the pulley system. This is due to friction in the bearings and also because energy must be supplied to raise the lower pulley. For example, if the weight of the lower pulley is 6 N, the potential energy gain each second by the lower pulley would be 0.9 J s⁻¹ (= 6 N × 0.15 m s⁻¹) when the load is raised at a speed of 0.15 m s⁻¹. Thus the energy wasted each second due to friction would be 0.6 J s⁻¹ (= 1.5 − 0.9 J s⁻¹).

4.4.2 Efficiency measures

Useful energy is energy transferred for a purpose. In any machine, where friction is present, some of the energy transferred by the machine is wasted. In other words, not all the energy supplied to the machine is transferred for the intended purpose. For example, suppose a 500 W electric winch raises a weight of 150 N by 6.0 m in 10 s:

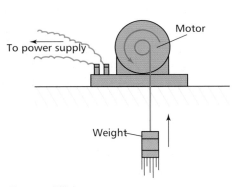

Fig 4.14 Efficiency

 • The electrical energy supplied to the winch is 500 W × 10 s = 5000 J

- The useful energy transferred by the machine is the potential energy gain of the load = 150 N × 6 m = 900 J.

Therefore, in this example, 4100 J of energy is wasted.

$$\text{The efficiency of a machine} = \frac{\text{useful energy transferred by the machine}}{\text{energy supplied to the machine}}$$

$$= \frac{\text{work done by the machine}}{\text{energy supplied to the machine}}$$

Notes

- **Percentage efficiency = efficiency × 100%**

In the above example, the efficiency of the machine is therefore 0.18 or 18%.

- Also: **Efficiency** $= \dfrac{\text{output power of a machine}}{\text{input power to the machine}}$

Wasting energy

In any process or device where energy is transferred for a purpose, the efficiency of the transfer process or the device is the fraction of the energy supplied which is used for the intended purpose. For example:

- A light bulb that is 10% efficient emits 10 J of energy as light for every 100 J of energy supplied to it by electricity. The rest of the energy is wasted as heat.

- An engine that is 45% efficient delivers 45 J of useful energy for every 100 J of energy supplied to it from its fuel. The rest of the energy is wasted as sound and heat.

Is it possible to stop energy being wasted as heat? In a power station, steam is used to drive turbines which turn the electricity generators. If the turbines were not kept cool, they would stop working because the pressure inside would build up and prevent steam entering. Stopping the heat transfer to the cooling water would stop the generators working. In general, energy tends to spread out when it is usefully used.

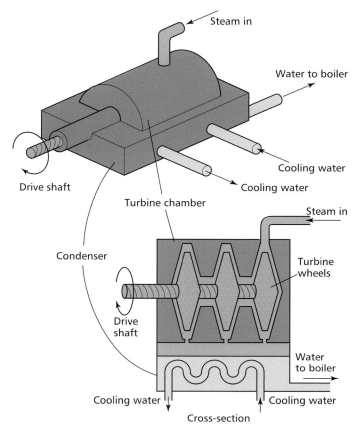

Fig 4.15 *A power station generator*

QUESTIONS

1 In a test of muscle efficiency, an athlete on an exercise bicycle pedals against a brake force of 30 N at a speed of 15 m s^{-1}.

  a Calculate the useful energy supplied per second by the athlete's muscles.

  b If the efficiency of the muscles is 25%, calculate the energy per second supplied to the athlete's muscles.

2 A 60 W electric motor raises a weight of 20 N through a height of 2.5 m in 8.0 s. Calculate:

  a the electrical energy supplied to the motor,

  b the useful energy transferred by the motor,

  c the efficiency of the motor.

3 A power station has an overall efficiency of 35% and it produces 200 MW of electrical power. The fuel used in the power station releases 80 MJ/kg of fuel burned. Calculate:

  a the energy per second supplied by the fuel,

  b the mass of fuel burned per day.

4 A vehicle engine has a power output of 6.2 kW and uses fuel which releases 45 MJ/kg when burned. At a speed of 30 m s^{-1} on a level road, the fuel usage of the vehicle is 18 km/kg. Calculate:

  a the time taken by the vehicle to travel 18 km at 30 m s^{-1},

  b the useful energy supplied by the engine in this time,

  c the overall efficiency of the engine.

1 a Calculate the kinetic energy of a 96 kg rugby player running at a speed of 6.0 m s^{-1}.

b What speed would a 2.4 kg rugby ball need to be moving at to have the same kinetic energy as the rugby player in part **a**?

c What would be the height gain of the 2.4 kg rugby ball in part **b**, if all its kinetic energy was converted into potential energy?

2 A brick, of mass 2.5 kg, falls off the top of a wall and hits the ground 2.5 m below.
Calculate:

a the loss of potential energy of the brick,

b the kinetic energy of the brick just before impact,

c the speed of the brick just before impact.

3 A ball, of mass 0.40 kg, released from a height of 1.2 m above a horizontal concrete floor hits the floor and rebounds to a maximum height of 0.80 m.

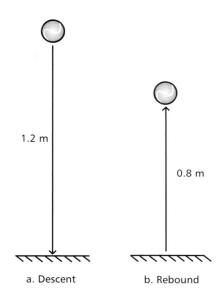

1.2 m

0.8 m

a. Descent   b. Rebound

a Calculate:
   (i) the loss of potential energy of the ball just before hitting the floor,
   (ii) the kinetic energy of the ball just before impact,
   (iii) the speed of the ball just before impact.

b Calculate:
   (i) the maximum gain of potential energy of the ball after impact,
   (ii) the kinetic energy of the ball just after impact,
   (iii) the speed of the ball just after impact.

4 An aircraft, of mass 2200 kg, takes off from an aircraft carrier and reaches a speed of 85 m s^{-1} in 20 s at a height of 320 m above the aircraft carrier.

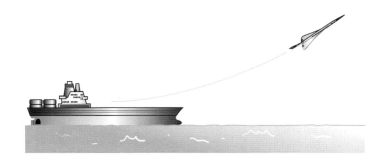

a Calculate:
   (i) the gain of potential energy of the aircraft,
   (ii) its gain of kinetic energy.

b (i) Estimate the average output power of the aircraft's engines in this time.
   (ii) If the aircraft's fuel has an energy value of 50 MJ kg^{-1} and the engines have an efficiency of 30%, estimate the mass of fuel burned in the 20 s taken to reach a height of 320 m.

5 A car, of mass 1200 kg, accelerates at a steady rate of 1.5 m s^{-2} along a straight level road.

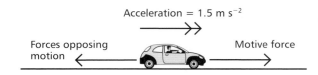

Acceleration = 1.5 m s^{-2}

Forces opposing motion   Motive force

a Calculate the resultant force acting on the car.

b The total force opposing the motion of the car is 500 N when the speed of the car is 7.0 m s^{-1}. Calculate the motive force of the engine at this speed when the acceleration is 1.5 m s^{-2}.

c Calculate:
   (i) the output power of the engine,
   (ii) the power wasted due to the forces opposing the motion of the car when the car's speed is 7.0 m s^{-1}.

d Account for the difference between your answers to parts **c**(i) and (ii).

6 An electric train is driven by an electric motor which has a maximum power output of 20 kW.

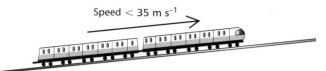

Speed < 35 m s⁻¹

a When the train is travelling at its maximum speed of 35 m s⁻¹ along a level track, its motor operates at full power. Calculate:
   (i) the motive force of the electric motor,
   (ii) the total force opposing the motion of the train at this speed.

b The train reaches an uphill incline, which causes its speed to decrease. In terms of energy changes, explain why the train speed becomes lower on the uphill incline than on the level track.

7 A tidal power station traps sea water at a high tide behind a barrier. When the tide has gone out, the trapped water is released through turbines that generate electricity.

a On a certain day, the trapped water covers an area of 5.0×10^7 m² at an average depth of 6.0 m above the turbines. Calculate:
   (i) the mass of this trapped water,
   (ii) the loss of potential energy of this trapped water when it is released through the turbines (the density of sea water = 1050 kg m⁻³).

b The trapped water in part **a**, is released in 4 hours through turbines that have an efficiency of 45%. Calculate the average power output of the turbines over the 4 hour release period.

8 An electric car is fitted with batteries which are used to drive an electric motor that provides the motive force for the car. The batteries have a maximum power output of 12 kW, which gives a maximum motive force of 600 N.

ELECTRA

a Calculate:
   (i) the top speed of the car,
   (ii) the car's maximum range, if the batteries last for 90 minutes at maximum power output without being recharged.

b Explain why the car would have a lower range if it stopped and started repeatedly instead of travelling at a constant speed.

Materials

Density

5.1.1 Density and its measurement

Lead is much more dense than aluminium. Sea water is more dense than tap water. To compare how much more dense one substance is compared with another, we need to measure the mass of equal volumes of the two substances. The substance with the greater mass in the same volume is more dense. For example, a lead sphere of volume 1 cm³ has a mass of 11.3 g; whereas an aluminium sphere of the same volume has a mass of 2.7 g.

> **The density of a substance is defined as its mass per unit volume.**

For a certain amount of a substance of mass m and volume V, its density, ρ (pronounced 'rho'), may be calculated using the equation:

$$\text{Density, } \rho = \frac{m}{V}$$

- The unit of density is the **kilogram per metre³ (kg m⁻³)**.
- Rearranging the above equation gives: $m = \rho V$

$$\text{or } V = \frac{m}{\rho}$$

More about units

Mass	1 kg = 1000 g
Length	1 m = 100 cm = 1000 mm
Volume	1 m³ = 10⁶ cm³
Density	$1000 \text{ kg m}^{-3} = \dfrac{10^6 \text{ g}}{10^6 \text{ cm}^3} = 1 \text{ g cm}^{-3}$

Table 5.1 shows the density of some common substances in kg m⁻³. You can see that gases are much less dense than solids or liquids.

Worked example

Using the data above, calculate:

a the mass, in kilograms, of a piece of aluminium of volume 3.6×10^{-5} m³,

b the volume, in m³, of a mass of 0.50 kg of iron.

Solution

a $\rho = 2700$ kg m⁻³; mass $m = \rho V = 2700$ kg m⁻³ $\times\ 3.6 \times 10^{-5}$ m³ $= 9.7 \times 10^{-2}$ kg

b $\rho = 7900$ kg m⁻³; volume $= \dfrac{m}{\rho} = \dfrac{0.50 \text{ kg}}{7900 \text{ kg m}^{-3}} = 6.3 \times 10^{-5}$ m³

Density measurements

An unknown substance can often be identified, if its density is measured and compared with the density of known substances. The following procedures may be used to measure the density of a substance.

- **A regular solid.** Measure its mass using a top-pan balance; measure its dimensions using a vernier or a micrometer. Calculate its volume using the appropriate formula (e.g. volume of a sphere of radius r is $\frac{4}{3}\pi r^3$; volume of a cylinder of radius r and length L is $\pi r^2 L$). Calculate the density from $\dfrac{\text{mass}}{\text{volume}}$.

Table 5.1 Density

Substance	Density/kg m⁻³
Air	1.2
Aluminium	2700
Copper	8900
Gold	19 300
Hydrogen	0.083
Iron	7900
Lead	11 300
Oxygen	1.3
Silver	10 500
Water	1000

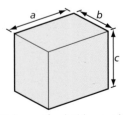

(i) Volume of cuboid = $a \times b \times c$

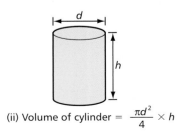

(ii) Volume of cylinder = $\dfrac{\pi d^2}{4} \times h$

Fig 5.1 *Volume formulae*

- **A liquid.** Measure the mass of an empty measuring cylinder. Fill the cylinder with the liquid and measure the volume of the liquid directly. Measure the mass of the cylinder and liquid to enable the mass of the liquid to be calculated. Calculate the density from $\dfrac{\text{mass of liquid}}{\text{volume}}$.

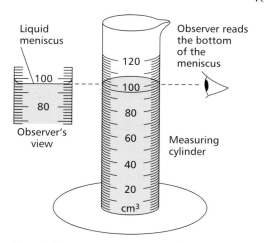

Fig 5.2 *Using a measuring cylinder*

- **An irregular solid.** Measure the mass of the object. Fill a displacement cap with water up to the spout. Place a beaker of known mass under the spout. Immerse the object on a thread in the liquid and collect the overflow. Measure the mass of the beaker and overflow. Hence determine the mass of the overflow water and calculate its volume, given the density of water is 1000 kg m^{-3}. Calculate the density of the object from

$$\dfrac{\text{its mass}}{\text{the overflow volume}}.$$

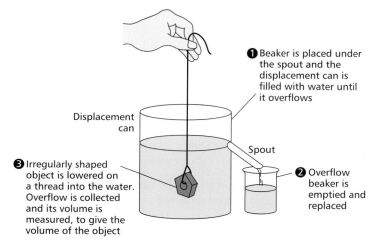

Fig 5.3 *Measuring the volume of an irregularly shaped object*

5.1.2 Density of alloys

An **alloy** is a solid mixture of two or more metals. For example, **brass** is an alloy of copper and zinc which has good resistance to corrosion and wear.

For an alloy that consists of two metals A and B, then for volume V of the alloy:

- If the volume of metal A is v_A, the mass of metal A is $\rho_A v_A$, where ρ_A is the density of metal A.
- If the volume of metal B is v_B, the mass of metal B = $\rho_B v_B$, where ρ_B is the density of metal B.

$\therefore$ **Mass of the alloy, $m = \rho_A v_A + \rho_B v_B$**

Hence the density of the alloy, $\rho = \dfrac{m}{V} = \dfrac{\rho_A v_A + \rho_B v_B}{V}$

$$= \dfrac{\rho_A v_A}{V} + \dfrac{\rho_B v_B}{V}$$

Worked example

A brass object consists of 3.3×10^{-5} m^3 of copper and 1.7×10^{-5} m^3 of zinc. Calculate the mass and the density of this object. The density of copper is 8900 kg m^{-3}; the density of zinc is 7100 kg m^{-3}.

Solution

Mass of copper   = density of copper $\times$ volume of copper
            = 8900 kg m$^{-3} \times 3.3 \times 10^{-5}$ m^3 = 0.29 kg

Mass of zinc   = density of zinc $\times$ volume of zinc
            = 7100 kg m$^{-3} \times 1.7 \times 10^{-5}$ m^3 = 0.12 kg

Total mass, m = 0.29 + 0.12 = 0.41 kg

Total volume, V = 5.0×10^{-5} m^3

Density of alloy, $\rho = \dfrac{m}{V} = \dfrac{0.41 \text{ kg}}{5.0 \times 10^{-5} \text{ m}^3} = 8200$ kg m^{-3}

QUESTIONS

1 A rectangular brick of dimensions 5.0 cm $\times$ 8.0 cm $\times$ 20.0 cm has a mass of 2.5 kg. Calculate:

 a its volume,

 b its density.

2 An empty paint tin of diameter 0.150 m and of height 0.120 m has a mass of 0.22 kg. It is filled with paint to within 7 mm of the top. Its total mass is then 6.50 kg. Calculate, for the paint in the tin:

 a the mass, **b** the volume, **c** the density.

3 A solid steel cylinder has a diameter of 12 mm and a length of 85 mm. Calculate:

 a its volume (in m^3),

 b its mass (in kg); the density of steel is 7800 kg m^{-3}.

4 An alloy tube of volume 1.8×10^{-4} m^3 consists, by volume, of 60% aluminium and 40% magnesium.

 a Calculate the mass, in the tube, of:
  (i) aluminium, (ii) magnesium.

 b Calculate the density of the alloy; the density of aluminium is 2700 kg m^{-3}; the density of magnesium is 1700 kg m^{-3}.

5.2.1 Pressure and force

Lots of people need to measure **pressure**. For example, nurses measure blood pressure, motor vehicle technicians measure tyre pressure and gas engineers measure the pressure of the gas supply. In these examples, a liquid or a gas inside a container presses on the container surface wherever the liquid or gas is in contact with the surface. An example of pressure due to a solid is where a brick rests on a surface. The weight of the brick presses down on the surface.

Pressure is defined as the force per unit area acting on a surface perpendicular to the surface.

- The pressure of a force F acting at right angles to a surface of area A is given by the equation:

$$\text{Pressure, } p = \frac{F}{A}$$

- The unit of pressure is the **pascal (Pa)**, which is equal to 1 N m^{-2}.

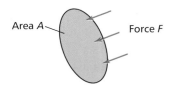

Area A — Force F

Fig 5.4 *Pressure*

Solid pressure

When a force acts on an object, the smaller the area over which the force acts, the greater the pressure of the force on a surface.

Liquid pressure

For any fluid at rest:

- the pressure at any point acts equally in all directions,
- the pressure increases with depth.

Gas pressure

The pressure of a gas is due to the gas molecules colliding elastically with the surface of the container. The molecules in a gas are in continuous random motion, colliding repeatedly with each other and the container surface. Each time a molecule hits the container surface, it exerts a small force on the surface. The effect of millions of molecules hitting the surface every second causes an even pressure to be exerted on the surface.

5.2.2 Hydraulics at work

A hydraulic system provides 'muscle power' enabling a large force to be exerted as a result of the action of a much smaller force. For example,

- **When a car tyre needs to be replaced,** a hydraulic car jack enables the wheel to be raised off the ground so it can be removed and the tyre on it replaced. When the effort is applied, a force acts on the piston in the narrow cylinder (Fig 5.6). The piston exerts pressure on the oil in the cylinder. The pressure causes a much larger force on the wide piston.

  For a force F_1 acting on a piston of cross-sectional area A_1, the pressure caused by this force is $\frac{F_1}{A_1}$. Therefore, the force exerted by this pressure on the wide piston is $\frac{F_1 A_2}{A_1}$, where A_2 is the cross-sectional area of the wide piston.

  Because $A_2 > A_1$, $F_2 > F_1$.

Fig 5.5 *A snow mobile*

Caterpillar tracks on snow vehicles allow the vehicle to travel across snow without sinking into the snow. The tracks have a much greater contact area with the ground than ordinary tyres. Therefore, the pressure of the tracks on snow is much less than the pressure of a tyre would be if the vehicle was fitted with tyres.

- **Vehicle brakes use hydraulics.** When the vehicle driver exerts a force on the brake pedal, pressure is exerted on the brake fluid in the master cylinder of the brake system. This pressure is transmitted along oil-filled brake pipes to a slave cylinder at each wheel (Fig 5.7). As a result, the piston in each slave cylinder pushes the brake pads onto the wheel. The disc pads are forced onto the wheel disc to slow the wheel down. The force on the brake pads is greater than the force applied to the piston in the master cylinder, because the area of cross-section of each slave cylinder is greater than the area of cross-section of the master cylinder.

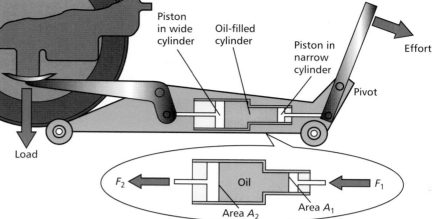

Fig 5.6 A hydraulic car jack

The efficiency of a hydraulic system

If air leaks into a hydraulic system, the pressure applied on the fluid in the system is not fully transmitted through the system. Some of the pressure is 'used' to compress the air in the system instead of acting on the slave piston. When a force F_1 acts on the master piston:

- The work done on the master piston = $F_1 s_1$, where s_1 is the distance moved by the master piston.

- The work done by each slave piston = $F_2 s_2$, where s_2 is the distance moved by each slave piston.

If the system has n slave cylinders,

its **efficiency** = $\dfrac{\text{work done by the system}}{\text{work done on the system}} = \dfrac{nF_2 s_2}{F_1 s_1}$

Because $F_2 = \dfrac{F_1 A_2}{A_1}$, then the efficiency = $\dfrac{nA_2 s_2}{A_1 s_1}$

where A_1 and A_2 are the cross-sectional areas of the master and slave cylinders, respectively.

The volume of fluid leaving the master cylinder = $A_1 s_1$ and the volume of fluid entering each slave cylinder = $A_2 s_2$.

Assuming the fluid in the system cannot be compressed, $A_1 s_1 = nA_2 s_2$ Therefore the system has an efficiency of 1 (= 100%) only if the fluid is incompressible. Oil with no air bubbles in is almost completely incompressible. However, the presence of air would reduce the efficiency considerably, as the volume of oil entering the slave cylinders would be less than the volume of oil leaving the master cylinder.

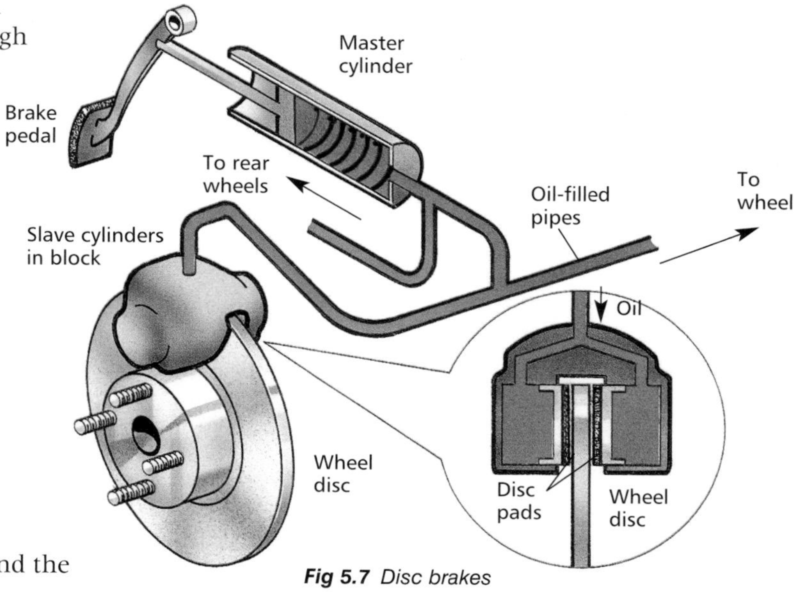

Fig 5.7 Disc brakes

QUESTIONS

$g = 9.8 \text{ m s}^{-2}$

1 Calculate the pressure exerted by a paving stone of density 2500 kg m^{-3} and of dimensions 0.80 m × 0.80 m × 0.05 m when the stone rests flat on a smooth horizontal surface.

2 Calculate the force due to the atmosphere on the panes of a sealed double-glazed window of dimensions 1.5 m × 0.80 m which has a vacuum between the two panes. The mean value of atmospheric pressure = 101 kPa.

3 The pressure in the tyres of a vehicle is 280 kPa. The contact area of each of the four tyres on the ground is 0.012 m². Calculate the weight of the vehicle.

4 A hydraulic brake system has a master cylinder of area of cross-section 5.0 × 10^{-4} m² and four slave cylinders, each of cross-sectional area 1.5 × 10^{-2} m². When a force of 120 N is applied to the master cylinder, calculate:

a (i) the pressure in the system,
   (ii) the force on each slave cylinder,

b the maximum distance moved by a slave piston, if the master piston moved a distance of 60 mm.

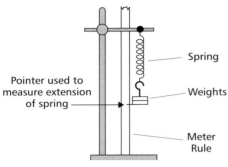

Pointer used to measure extension of spring →

Spring

Weights

Meter Rule

a Testing the extension of a spring

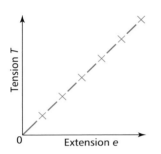

Tension T

Extension e

b Hooke's Law

Fig 5.8

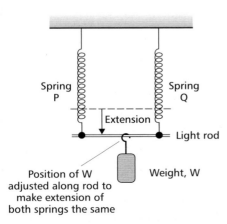

Spring P

Spring Q

Extension

Light rod

Position of W adjusted along rod to make extension of both springs the same

Weight, W

Fig 5.9 Two springs in parallel

5.3 Springs

5.3.1 Hooke's Law

A stretched spring exerts a pull on the object holding each end of the spring. This pull, referred to as the **tension** in the spring, is equal and opposite to the force needed to stretch the spring. The more a spring is stretched, the greater the tension in it. Figure 5.8a shows a stretched spring supporting some weights at rest. This arrangement may be used to investigate how the tension in a spring depends on its extension from its unstretched length. The measurements may be plotted on a graph of tension v. extension (Fig 5.8b). The graph shows that the force needed to stretch a spring is proportional to the extension of the spring. This is known as **Hooke's Law**, after Robert Hooke, a 17th-century scientist.

> **Hooke's Law states that the force needed to stretch a spring is proportional to the extension of the spring from its natural length.**

Hooke's Law may be written as:

$$\textbf{Force, } F = ke$$

where k is the spring constant (sometimes referred to as the stiffness constant) and e is the extension.

- The greater the value of k, the stiffer the spring is. The unit of k is N m^{-1}.
- The graph of F against e is a straight line of gradient k through the origin.
- If a spring is stretched beyond its **elastic limit**, it does not regain its initial length when the force applied to it is removed.
- AS/A level maths students may meet Hooke's Law in the form $F = \dfrac{\lambda e}{L}$ where L is the unstretched length of the spring and λ $(= kL)$ is the spring modulus. λ is not in the specification for AS/A level physics.

Worked example

A vertical steel spring fixed at its upper end has an unstretched length of 300 mm. Its length is increased to 385 mm when a 5.0 N weight attached to the lower end is at rest. Calculate:

a the spring constant,

b the length of the spring when it supports an 8.0 N weight at rest.

Solution

a Use $F = ke$ with $F = 5.0$ N and $e = 385 - 300$ mm $= 85$ mm $= 0.085$ m.

Therefore $k = \dfrac{F}{e} = \dfrac{5.0 \text{ N}}{0.085 \text{ m}} = 59$ N m^{-1}

b Use $F = ke$ with $F = 8.0$ N and $k = 59$ N m^{-1} to calculate e:

$$e = \dfrac{F}{k} = \dfrac{8.0 \text{ N}}{59 \text{ N m}^{-1}} = 0.136 \text{ m}$$

Therefore the length of the spring $= 0.300$ m $+ 0.136$ m $= 0.436$ m

Springs in parallel

Figure 5.9 shows a weight W supported by means of two springs, P and Q, in parallel with each other. The extension, e, of each spring is the same. Therefore:

- The force needed to stretch P, $F_P = k_P e$
- The force needed to stretch Q, $F_Q = k_Q e$, where k_P and k_Q are the spring constants of P and Q respectively.

The weight W is supported by both springs, $W = F_P + F_Q = k_P e + k_Q e = k_{eff} e$ where the **effective spring constant,** $k_{eff} = k_P + k_Q$

Springs in series

Figure 5.10 shows a weight W supported by means of two springs joined end-on in 'series' with each other. The tension in each spring is the same and is equal to the weight W. Therefore:

- Extension of spring P, $e_P = \dfrac{W}{k_P}$

- Extension of spring Q, $e_Q = \dfrac{W}{k_Q}$, where k_P and k_Q are the spring constants of P and Q respectively.

$$\textbf{Total extension, } e = e_P + e_Q = \frac{W}{k_P} + \frac{W}{k_Q} = \frac{W}{k_{eff}}$$

where k_{eff}, the effective spring constant, is given by the equation $\dfrac{1}{k_{eff}} = \dfrac{1}{k_P} + \dfrac{1}{k_Q}$

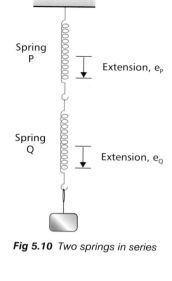

Spring P — Extension, e_P

Spring Q — Extension, e_Q

Fig 5.10 *Two springs in series*

5.3.2 The energy stored in a stretched spring

Elastic potential energy is stored in a stretched spring. If the spring is suddenly released, the elastic energy stored in it is suddenly converted to kinetic energy of the spring. As explained in section 4.1, the work done to stretch a spring by extension e_0 from its unstretched length is $\frac{1}{2}F_0 e_0$, where F_0 is the force needed to stretch the spring to extension e_0. The work done on the spring is stored as elastic potential energy. Therefore the elastic potential energy E_p in the spring is $\frac{1}{2}F_0 e_0$. Also, since $F_0 = ke_0$, where k is the spring constant, then $E_p = \frac{1}{2}ke_0^2$.

Elastic potential energy stored in a stretched spring, $E_p = \frac{1}{2}F_0 e_0 = \frac{1}{2}ke_0^2$

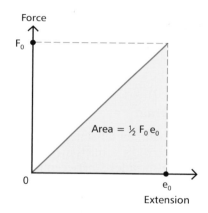

Force

F_0

Area = ½ F_0 e_0

0   e_0

Extension

Fig 5.11 *Energy stored in a stretched spring*

QUESTIONS

$g = 9.8 \text{ m s}^{-2}$

1 A steel spring has a spring constant of 25 N m⁻¹. Calculate:

  a the extension of the spring when the tension in it is equal to 10 N,

  b the tension in the spring when it is extended by 0.50 m from its unstretched length.

2 Two identical steel springs of length 250 mm are suspended vertically side-by-side from a fixed point. A 40 N weight is attached to the ends of the two springs. The length of each spring is then 350 mm. Calculate:

  a the tension in each spring,

  b the extension of each spring,

  c the spring constant of each spring.

3 Repeat question **2a** and **b** for the two springs in 'series' and vertical.

4 An object of mass 0.150 kg is attached to the lower end of a vertical spring of unstretched length 300 mm, which is fixed at its upper end. With the object at rest, the length of the spring becomes 420 mm as a result. Calculate:

  a the spring constant,

  b the energy stored in the spring,

  c the weight that needs to be added to extend the spring to 600 mm.

5.4 Deformation of solids

5.4.1 Force and solid materials

Look around at different materials and think about the effect of force on each material. To stretch, or twist, or compress the material, a pair of forces is needed. For example, stretching a rubber band requires the rubber band to be pulled by a force at either end. Some materials, such as rubber, bend or stretch easily. The **elasticity** of a solid material is its ability to regain its shape after it has been deformed or distorted, and the forces that deformed it have been released. Deformation that stretches an object is **tensile**, whereas deformation that compresses an object is **compressive**.

The arrangement shown in Fig 5.8 **a** p.72 may be used to test different materials to see how easily they stretch. In each case, the material is held at its upper end and loaded by hanging weights at its lower end. The position of the pointer is measured as the weight of the load is increased in steps then decreased to zero. The extension of the strip of material at each step is its increase of length from its unloaded length. The tension in the material is equal to the weight. The measurements may be plotted as a graph of tension v. extension, as shown in Figure 5.12.

- A steel spring gives a straight line, in accordance with Hooke's Law (see section 5.3).
- A rubber band at first extends easily when it is stretched. However, it becomes 'fully stretched' and very difficult to stretch further when it has been lengthened considerably.
- A polythene strip 'gives' and stretches easily after its initial stiffness is overcome. However, after 'giving', it extends little and becomes difficult to stretch.

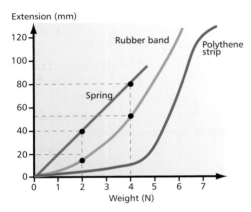

Fig 5.12 Typical curves

5.4.2 Stress and strain

The extension of a wire under tension may be measured using Searle's apparatus (Fig 5.13). A micrometer attached to the control wire is adjusted so the spirit level between the control and test wire is horizontal. When the test wire is loaded, it extends slightly causing the spirit level to drop on one side. The micrometer is then readjusted to make the spirit level horizontal again. The change of the micrometer reading is therefore equal to the extension. The extension may be measured for different values of tension by increasing the test weight in steps.

For a wire of length L and area of cross-section A under tension:

- The **tensile stress** in the wire, $\sigma = \dfrac{F}{A}$, where F is the tension. The unit of stress is the **pascal (Pa)**, equal to $1\ \text{N m}^{-2}$.

- The **tensile strain** in the wire, $\varepsilon = \dfrac{e}{L}$, where e is the extension of the wire (i.e. change of length, ΔL). Strain is a ratio and therefore has **no unit**.

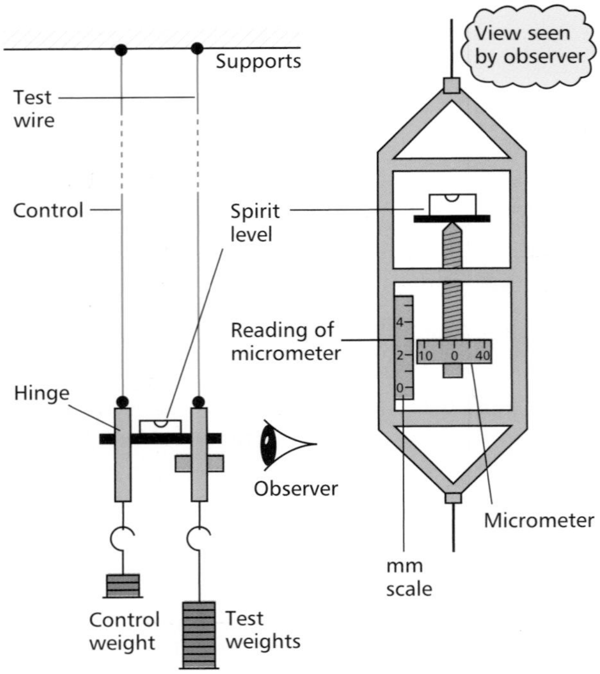

Fig 5.13 Searle's apparatus

Stress and strain graphs

Figure 5.14 shows how the stress in a wire varies with strain.

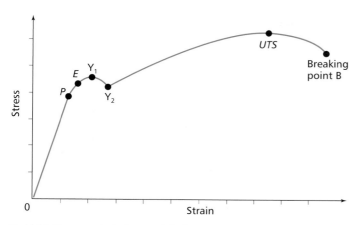

Fig 5.14 *Stress v. strain for a metal wire*

- **From 0 to the limit of proportionality P**, the stress is proportional to the strain.

  The value of stress/strain is a constant, known as the **Young modulus** of the material:

  $$\textbf{Young modulus, } E = \frac{\text{stress, } \sigma}{\text{strain, } \varepsilon} = \frac{\frac{F}{A}}{\frac{e}{L}} = \frac{FL}{Ae}$$

  Note that the area of cross-section of a wire, of uniform diameter d, $= \frac{\pi d^2}{4}$. Also, the unit of E is the pascal (Pa).

- **Beyond P**, the line curves and continues beyond the **elastic limit** E to the **yield point** Y_1, which is where the wire weakens temporarily. The elastic limit is the point beyond which the wire is permanently stretched and suffers **plastic deformation**.

- **Beyond Y_2**, a small increase in the stress causes a large increase in strain as the material of the wire undergoes **plastic flow**. Beyond maximum stress, or the **Ultimate Tensile Stress** (UTS), the wire loses its strength, extends and becomes narrower at its weakest point. Increase of stress occurs at this point due to the reduced area of cross-section, until the wire breaks at B. The stress at B is called the **breaking stress**.

Stress and strain curves for different materials

- The **stiffness** of different materials can be compared using the **gradient** of the stress-strain line which is equal to the Young modulus of the material. Thus steel is stiffer than copper (Fig 5.15).

- The **strength** of a material is its Ultimate Tensile Stress (UTS), which is its maximum stress. Steel is stronger than copper because its maximum stress is greater.

- A **brittle** material snaps without any noticable yield. For example, glass breaks without any 'give'.

- A **ductile** material can be drawn into a wire. Copper is more ductile than steel.

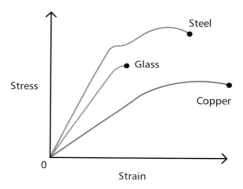

Fig 5.15 *Stress-strain curves*

QUESTIONS

1 Calculate the stress in a wire of diameter 0.25 mm when the tension in the wire is 50 N.

2 A metal wire of diameter 0.23 mm and of unstretched length 1.405 m was suspended vertically from a fixed point. When a 40 N weight was suspended from the lower end of the wire, the wire stretched by an extension of 10.5 mm. Calculate the Young modulus of the wire material.

3 A vertical steel wire of length 2.5 m and diameter 0.35 mm supports a weight of 90 N. Calculate:

  a the stress in the wire,

  b the extension of the wire; Young modulus of steel $= 2.0 \times 10^{11}$ Pa.

4 Compare the stress-strain curves in Figure 5.16. Use the curves to decide:

  a which material is: (i) stiffest, (ii) strongest?

  b which material is: (i) brittle, (ii) ductile?

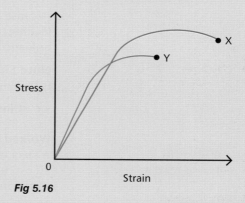

Fig 5.16

More about stress and strain

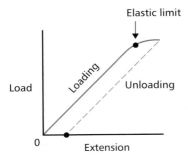

a *Metal wire*

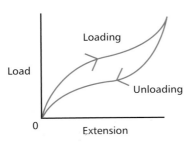

b *Rubber*

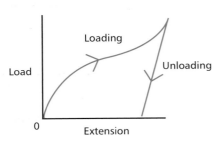

c *Polythene*

Fig 5.17 *Loading and unloading curves*

5.5.1 Investigating loading and unloading of different materials

How does the **strength** of a material change as a result of being stretched? To investigate this question, the tension in a strip of material is increased by increasing the weight it supports in steps. At each step, the extension of the material is measured. Typical results for different materials are shown in Figure 5.17. For each material, the loading curve and the subsequent unloading curve are shown.

- For a metal wire, its extension at any given tension when it is being unloaded is exactly the same as when it was being loaded (Fig 5.17a). The unloading curve and the loading curve are the same straight line. Provided its elastic limit is not exceeded, the wire returns to the same length when it has been completely unloaded as it had before it was loaded. If the wire is stretched beyond its elastic limit, the unloading line is parallel to the loading line. In this case, when it is completely unloaded, the wire is slightly longer so it has a **permanent extension**.

- For a rubber band (Fig 5.17b), the extension during unloading at any given tension is greater than during loading. The rubber band returns to the same unstretched length, but the unloading curve is below the loading curve except at zero extension and maximum extension. The rubber band remains elastic as it regains its initial length, but it has a **low limit of proportionality**.

- For a polythene strip (Fig 5.17c), the extension during unloading is also greater than during loading. However, the strip does not return to the same initial length when it is completely unloaded. The polythene strip has a low limit of proportionality and suffers **plastic deformation**.

5.5.2 Strain energy

As explained in section 5.3, the area under a graph of force v. extension is equal to the work done to stretch the wire. The work done to deform an object is referred to as **strain energy**. Consider the energy changes for each of the three materials above when each material is loaded then unloaded.

Metal wire (or spring)

Provided the limit of proportionality is not exceeded, the work done to stretch a wire to extension e is $\frac{1}{2}Fe$, where F is the tension in the wire at this extension (see section 5.3). Because the elastic limit is not reached, the work done is stored as elastic energy in the wire. Therefore:

$$\textbf{Elastic energy stored in a stretched wire} = \tfrac{1}{2}Fe$$

Because the graph of tension against extension is the same for unloading as for loading, all the energy stored in the wire can be recovered when the wire is unloaded.

Note Since the volume of the wire $= AL$, where A is its cross-sectional area and L is its length, then:

$$\textbf{Elastic energy stored per unit volume} = \tfrac{1}{2}\frac{Fe}{AL} = \tfrac{1}{2} \times \textbf{stress} \times \textbf{strain}$$

Worked example

A steel wire of uniform diameter 0.35 mm and of length 810 mm is stretched to an extension of 2.5 mm. Calculate: **a** the tension in the wire, **b** the elastic energy stored in the wire.

The Young modulus for steel $= 2.1 \times 10^{11}$ Pa

Solution

a Extension, $e = 2.5$ mm $= 2.5 \times 10^{-3}$ m

Area of cross-section of wire $= \dfrac{\pi (0.35 \times 10^{-3})^2}{4} = 9.6 \times 10^{-8}$ m²

To find the tension, rearranging the Young modulus equation $E = \dfrac{FL}{Ae}$ gives

$$F = \dfrac{EAe}{L} = \dfrac{2.1 \times 10^{11} \times 9.6 \times 10^{-8} \times 2.5 \times 10^{-3}}{0.810} = 62 \text{ N}$$

a Elastic energy stored in the wire $= \frac{1}{2}Fe = 0.5 \times 62 \times 2.5 \times 10^{-3} = 7.8 \times 10^{-2}$ J

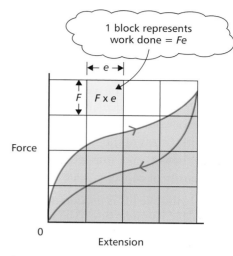

1 block represents work done = Fe

Fig 5.18 *Energy changes when loading and unloading rubber*

Rubber band

The work done to stretch the rubber band is represented in Figure 5.17b by the area under the loading curve. The work done by the rubber band when it is unloaded is represented by the area under the unloading curve. The area between the loading curve and the unloading curve therefore represents the difference between energy stored in the rubber band when it is stretched and useful energy recovered from it when it unstretches. The difference is because some of the energy stored in the rubber band becomes internal energy of the molecules when the rubber band unstretches. Figure 5.18 shows how the area between the loading and unloading curve can be used to determine the internal energy retained when the rubber band is stretched then released.

Polythene

As it does not regain its initial length, the area between the loading and unloading curve represents work done to deform the material permanently, as well as internal energy retained by the polythene when it unstretches.

The plastic behaviour of polythene is because polythene is a polymer, so each molecule is a long chain of atoms. Before being stretched, the molecules are tangled together or folded against each other. Weak bonds referred to as **cross-links** form between atoms where molecules are in contact with each other. When placed under tension, a thin sample of polythene easily stretches as the original weak cross-links break and the molecules align parallel to each other. New weak cross-links form in the stretched state and, when the tension is removed, the polythene strip remains stretched.

Note Rubber is also a polymer, but its molecules are curled up and tangled together when it is unstretched. When placed under tension, its molecules are straightened out as its length increases more and more. When the tension is removed, its molecules curl up again and it regains its initial length.

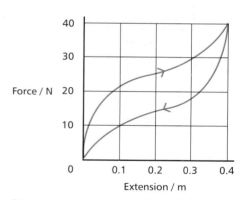

Fig 5.19 *Force v. extension for a strip of rubber*

QUESTIONS

Young modulus for steel $= 2.1 \times 10^{11}$ Pa; Young modulus for copper $= 1.3 \times 10^{11}$ Pa

1 A vertical steel cable of diameter 24 mm and of length 18 m supports a weight of 1500 N attached to its lower end. Calculate:

  a the tensile stress in the cable,

  b the extension of the cable,

  c the elastic energy stored in the cable, assuming its elastic limit has not been reached.

2 A vertical steel wire of diameter 0.28 mm and of length 2.0 m is fixed at its upper end and has a weight of 15 N suspended from its lower end. Calculate:

  a the extension of the wire,

  b the elastic energy stored in the wire.

3 A steel bar of length 40 mm and cross-sectional area 4.5×10^{-4} m² is placed in a vice, and compressed by 0.20 mm when the vice is tightened. Calculate:

  a the compressive force exerted on the bar,

  b the work done to compress it.

4 Figure 5.19 shows a force v. extension curve for a strip of rubber. Use the graph to determine:

  a the work done to stretch the rubber to an extension of 0.40 m,

  b the internal energy retained by the rubber when it unstretches.

1 A uniform copper wire of length 2.50 m has a diameter of 0.32 mm. Calculate:
  a the volume of the wire,
  b the mass of the wire,
  c the mass per unit length of the wire.
  Density of copper = 8900 kg m^{-3}

2 A hydraulic lift is used in a garage to raise a vehicle to enable its underside to be repaired. The figure below shows how such a lift works. The maximum pressure of the fluid in the hydraulic lift is 1.2 MPa.

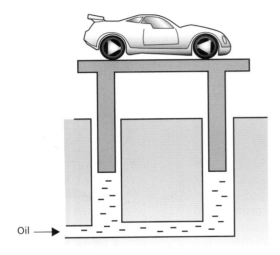

Oil →

  a The area of cross-section of each of the four 'legs' of the lift is 9.0×10^{-3} m². Calculate the maximum safe load the lift can raise if the lift platform has a weight of 1.5×10^4 N.
  b The lift is used to raise a vehicle of mass 2200 kg through a height of 2.4 m.
    (i) Calculate the gain of potential energy of the vehicle.
    (ii) Calculate the work done by the hydraulic system to lift the vehicle by 2.4 m.
    (iii) Account for the difference between your answer to (i) and (ii).

3 A steel spring of length 300 mm fixed at its upper end hangs vertically. When a 4.0 N weight is suspended from its lower end, the spring extends to an equilibrium length of 420 mm.

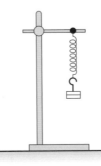

  a Calculate:
    (i) the spring constant of this spring,
    (ii) the length of the spring when it supports a weight of 6.0 N.
  b A second identical spring is suspended from the lower end of the first spring. Calculate the weight that would need to be suspended on the end of the lower spring to give a total extension for both springs equal to 90 mm.

4 a With the aid of a suitable example, explain what is meant by:
    (i) a brittle material,
    (ii) a ductile material.
  b A steel guitar wire of diameter 0.28 mm and of length 800 mm is tightened by turning its tension key by 2 turns, each turn increasing the length of the wire by 1.8 mm. Calculate the increase of tension in the wire as a result.
  Young modulus of steel = 2.1×10^{11} N m^{-2}

5 a Explain what is meant by the following terms:
    (i) the elastic limit of a strip of material,
    (ii) plastic behaviour.
  b (i) Sketch a graph to show how the extension of a rubber band varies with the force used to stretch it.
    (ii) When a car is in motion, each part of its tyres that make contact with the road is squashed and stretched as it passes through the point of contact with the road. In energy terms, explain why this repeated squashing and stretching causes the tyre to become warm.

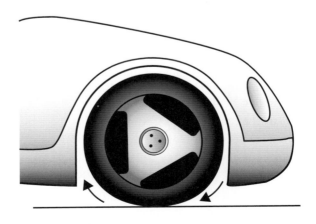

6 a Define the Young modulus of a material.

b The figure below shows stress-strain curves for three different metals, X, Y and Z.
  (i) State which of the three metals is strongest, giving your reason for this choice.
  (ii) State which of the three metals is most ductile, giving a reason for this choice.

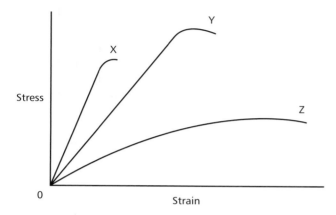

7 A crane is designed to lift a load of maximum weight of 9000 N to a maximum height of 65 m. The crane cable has an area of cross-section of 4.5×10^{-4} m² and is attached to a pulley block and hook of weight 600 N, so that the load is supported by two lengths of the crane cable (as shown in the figure below).

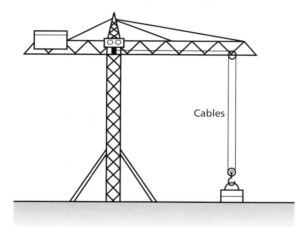

Cables

a Calculate the maximum weight of the suspended crane cable, pulley block and hook when the hook is just off the ground.

b (i) Calculate the increase of stress in the crane cables when a load of 9000 N is lifted off the ground.
  (ii) Show that the extension of each of the two lengths of the crane cable when the load is raised off the ground is 3.3 mm. Density of steel = 8000 kg m⁻³; Young modulus of steel = 2.0×10^{11} N m⁻²

8 The figure below shows a graph of the force used to stretch a metal wire against the extension of the wire for a wire of length 1620 mm and of cross-sectional area 1.40×10^{-7} m².

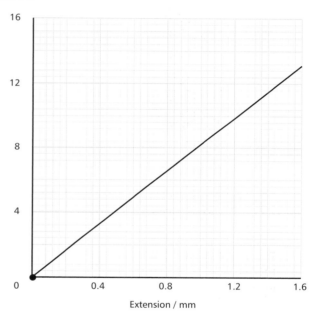

a Calculate the Young modulus of the wire.

b Calculate the energy stored in the wire when it extended from its unstretched length by 1.60 mm.

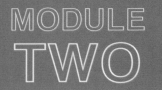

MODULE TWO

Electrons and photons

6

Electric current

6.1

Current and charge

6.1.1 Electrical conduction

An **electric current** is a flow of charge due to the passage of charged particles. These charged particles are referred to as **charge carriers**.

- In metals, the charge carriers are **conduction electrons**. They move about inside the metal, repeatedly colliding with each other and the fixed positive ions in the metal.
- In comparison, when an electric current is passed through a salt solution, the charge is carried by **ions** which are charged atoms or molecules.

A simple test for conduction of electricity is shown in Figure 6.1. If the test material is a metal, electrons pass round the circuit: leaving the battery at its negative terminal, then passing though the metal, then re-entering the battery at its positive terminal.

The convention for the direction of current in a circuit is from + **to** −, as shown in Figure 6.2. The convention was agreed long before the discovery of electrons. When it was set up, it was known that an electric current is a flow of charge one way round a circuit. However, it was not known if the current was due to positive charge flowing round the circuit from + to −, or if it was due to negative charge flowing from − to +.

- The unit of current is the **ampere (A)**, which is defined in terms of the force between two parallel wires when they pass the same current. See section 8.3.
- The unit of charge is the **coulomb (C)**, equal to the charge flow in one second when the current is one ampere. The magnitude of the charge of the electron, e, is 1.6×10^{-19} C. This is sometimes referred to as the **elementary charge**.
- For a current I, the **charge flow** ΔQ in time Δt is given by the equation:

$$\Delta Q = I\Delta t$$

- For charge flow ΔQ in a time interval Δt, the current I is given by:

$$I = \frac{\Delta Q}{\Delta t}$$

The equation shows that a current of 1 A is due to a flow of charge of 1 coulomb per second. As the charge of the electron is 1.6×10^{-19} C, a current of 1 A along a wire must be due to 6.25×10^{18} electrons passing along the wire each second.

6.1.2 More about charge carriers

Conductors, insulators and semiconductors
Materials can be classified in electrical terms as either conductors, insulators or semiconductors.

- In an **insulator**, each electron is attached to an atom and cannot move away from the atom. When a voltage is applied across an insulator, no current passes through the insulator because no electrons can move through the insulator.

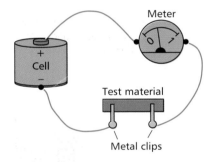

Fig 6.1 *Testing for conduction*

Fig 6.2 *Convention for current*

- In a **metallic conductor**, most electrons are attached to atoms; but some are not and these are the **charge carriers** in the metal. When a voltage is applied across the metal, these **conduction electrons** move towards the positive terminal of the metal.

- In a **semiconductor**, the number of charge carriers increases with increase of temperature. The resistance of a semiconductor therefore decreases as its temperature is raised. A pure semiconducting material is referred to as an **intrinsic semiconductor**, because conduction is due to electrons that break free from the atoms of the semiconductor.

- In an **electrolyte**, the charge carriers are **positive and negative ions**. When a voltage is applied to the electrodes, the positive ions are attracted to the **cathode** (i.e. the negative electrode) and the negative ions are attracted to the **anode** (i.e. the positive electrode). The ions are discharged on reaching the relevant electrode.

An electrolysis experiment

A metal object can be plated with copper by using it as the cathode in an electrolyte consisting of copper sulphate solution. Copper ions carry a positive charge, so they are attracted to the cathode where they are discharged.

The mass of copper deposited at the cathode depends on the **current** and the **time** for which the current is on. By weighing the cathode before and after plating, it is possible to measure the mass of copper deposited on the cathode. Measurements show that the mass of copper deposited is proportional to the current × the time taken, and therefore to the **charge** passed through the solution. This is because each copper ion discharged at the cathode gains **two electrons** from the cathode. So, the total number of copper ions discharged is proportional to the total number of electrons passed.

Physics and the Human Genome Project

To map the human genome, fragments of DNA are tagged with a C, or G, or A, or T base containing a dye. Each tagged fragment carries a negative charge. A voltage is applied across a strip of gel containing a spot of liquid containing tagged fragments. The fragments are attracted to the positive electrode. The smaller the fragment, the faster it moves; so the fragments separate out according to size, as they move to the positive electrode. The fragments pass through a spot of laser light, which causes the dye attached to each tag to fluoresce as it passes through the laser spot. Light sensors linked to a computer detect the glow from each tag. The computer is programmed to work out and display the sequence of bases in the DNA fragments.

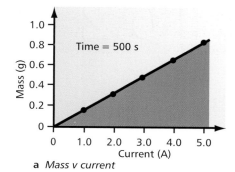

a *Mass v current*

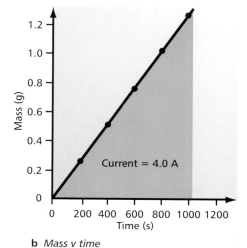

b *Mass v time*

Fig 6.3 *Copper plating*

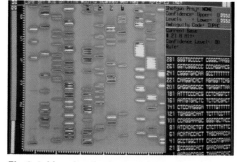

Fig 6.4 *Mapping the human genome*

QUESTIONS

$e = 1.6 \times 10^{-19}$ C

1 a The current in a certain wire is 0.35 A. Calculate the charge passing a point in the wire:
(i) in 10 s,   (ii) in 10 min.

b Calculate the average current in a wire through which a charge of 15 C passes in:
(i) 5 s,   (ii) 100 s.

2 Calculate the number of electrons passing a point in the wire in 1 min when the current is:

a 1 μA   **b** 5.0 A

3 a Use Figure 6.3 to show that 3.3×10^{-7} kg of copper is deposited in a copper plating experiment when a charge of 1 C is passed.

b The charge carried by a copper ion is +2e. Use the information in part **a** to calculate the mass of a copper ion.

4 A certain type of rechargeable battery is capable of delivering a current of 0.2 A for 4000 s, before its voltage drops and it needs to be recharged. Calculate:

a the total charge the battery can deliver before it needs to be recharged,

b the maximum time it could be used for without being recharged if the current through it were:
(i) 0.5 A   (ii) 0.1 A

6.2.1 Energy and potential difference

When a torch bulb is connected to a battery, electrons deliver energy from the battery to the torch bulb. Each electron which passes through the bulb takes a fixed amount of energy from the battery and delivers it to the bulb. After delivering energy to the bulb, each electron re-enters the battery via the positive terminal to be resupplied with more energy to deliver to the bulb.

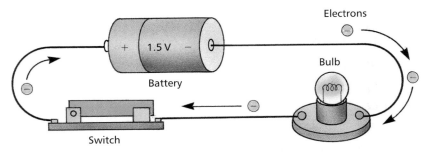

Fig 6.5 *Energy transfer by electrons*

Each electron in the battery has the **potential** to deliver energy, even if the battery is not part of a complete circuit. In other words, the battery supplies each electron with **electrical potential energy**. When the battery is in a circuit, each electron passing through a circuit component does work to pass through the component and therefore uses some or all of its electrical potential energy. The **work done** by an electron is equal to its loss of potential energy. The work done per unit charge is defined as the **potential difference** (abbreviated as p.d.) or **voltage, *V*,** across the component.

> **Potential difference is defined as the work done (or energy transfer) per unit charge.**

The unit of p.d. is the **volt (V),** which is equal to 1 joule per coulomb.

If work *W* is done when charge *Q* flows through the component, the p.d. across the component, *V*, is given by:

$$V = \frac{W}{Q}$$

Rearranging this equation gives $W = QV$ for the work done or energy transfer when charge *Q* passes through a component which has a p.d. *V* across its terminals.

Examples

- If 30 J of work is done when 5 C of charge passes through a component, the p.d. across the component must be 6 V $\left(= \dfrac{30\ \text{J}}{5\ \text{C}} \right)$.

- If the p.d. across a component in a circuit is 12 V, then 3 C of charge passing through the component would transfer 36 J of energy from the battery to the component.

Energy transfer in different devices

An electric current has a **heating effect** when it passes through a component with **resistance**. It also has a **magnetic effect**, which is made use of in electric motors and loudspeakers.

Fig 6.6 *Electrical devices*

- In a device that has resistance, such as an electrical heater, the work done on the device is transferred as **thermal energy**. This happens because the charge carriers repeatedly collide with atoms in the device and transfer energy to them, so the atoms **vibrate more** and the resistor becomes hotter.
- In an electric motor, the work done on the motor is transferred as **kinetic energy** of the motor. The charge carriers are electrons that need to be forced through the wires of the **spinning motor coil**, against the opposing force on the electrons due to the motor's magnetic field.
- For a loudspeaker, the work done on the loudspeaker is transferred as **sound energy**. Electrons need to be forced through the wires of the **vibrating loudspeaker coil**, against the force on them due to the loudspeaker magnet.

Electromotive force (e.m.f)

> **The e.m.f., ε, of a source of electricity is defined as the electrical energy per unit charge produced inside the source.**

The unit of e.m.f. is the **volt (V)**, the same as the unit of p.d., equal to 1 joule per coulomb.

For a source of e.m.f., ε, in a circuit, the electrical energy produced when charge Q passes through the source is $Q\varepsilon$. This energy is transferred to other parts of the circuit, and some may be dissipated in the source itself due to the source's internal resistance (see section 7.3 for internal resistance).

Fig 6.7 *Power supplies*

6.2.2 Electrical power and current

Consider a component, or device, which has a potential difference V across its terminals and a current I passing through it. In time Δt:

- The charge flowing through it, $Q = I\Delta t$
- The work done by the charge carriers, $W = QV = (I\Delta t)V = IV\Delta t$

Work done, $W = IV\Delta t$

The energy transfer ΔE in the component or device is equal to the work done W. Therefore, because power $= \dfrac{\text{energy}}{\text{time}}$, the electrical power P supplied to the device is given by:

$$P = \frac{IV\Delta t}{\Delta t} = IV$$

Electrical power, $P = IV$

Notes

- This equation can be rearranged to give: $I = \dfrac{P}{V}$ or $V = \dfrac{P}{I}$
- The unit of power is the **watt (W)**. Therefore one volt is equal to one watt per ampere. For example, if the p.d. across a component is 4 V, then the power delivered to the component is 4 W per ampere of current.
- Energy supplied by mains electricity is measured in **kilowatt hours (kW h)**, usually referred to as 'units'. One kilowatt hour is the energy transfer when 1 kW of power is supplied for exactly 1 h.

  Therefore 1 kW h = 3.6 MJ ($= 1000$ W $\times$ 3600 s).
- The correct value of a **fuse** for an electrical appliance can be worked out using the equation $I = \dfrac{P}{V}$ if the power and voltage of the appliance are known.

  For example, a 230 V, 2.5 kW electric kettle would take a current of 11 A ($= \dfrac{2500\text{ W}}{230\text{ V}}$). Therefore, given the choice between a 5 A fuse or a 13 A fuse, a 13 A fuse should be used.

Worked example

A 12 V, 48 W electric heater is connected to a 12 V battery. Calculate:

a the heater current, **b** the energy transfer in 300 s.

Solution

a Rearrange $P = IV$ to give $I = \dfrac{P}{V} = \dfrac{48\text{ W}}{12\text{ V}} = 4$ A

b $\Delta E = IV\Delta t = 4$ A $\times$ 12 V $\times$ 300 s $= 14\,400$ J

QUESTIONS

1 Calculate the energy transfer in 1200 s in a component when the p.d. across it is 12 V and the current is:

  a 2 A   **b** 0.05 A

2 A 6 V, 12 W light bulb is connected to a 6 V battery. Calculate:

  a the current through the light bulb,

  b the energy transfer to the light bulb in 1800 s.

3 A 230 V electrical appliance has a power rating of 800 W.

  a Calculate:

    (i) the energy transfer in the appliance in 1 min,

    (ii) the current taken by the appliance.

  b Which of the following fuse values would be suitable for this appliance: 3 A, 5 A, or 13 A?

4 A battery has an e.m.f. of 9 V and a negligible internal resistance. It is capable of delivering a total charge of 1350 C. Calculate:

  a the maximum energy the battery could deliver,

  b the power it would deliver to the components of a circuit if the current through it was 0.5 A,

  c how long the battery would last for if it was to supply power at the rate calculated in part **b**.

6.3 Resistance

6.3.1 Definitions and laws

The **resistance** of a component in a circuit is a measure of the difficulty of making current pass through the component. Resistance is caused by the repeated collisions between the charge carriers in the material with each other and with the fixed positive ions of the material.

$$\text{Resistance of any component} = \frac{\text{p.d. across the component}}{\text{current through it}}$$

For a component which passes current I when the p.d. across it is V, its resistance R is given by the equation:

$$R = \frac{V}{I}$$

The unit of resistance is the **ohm (Ω)** which is equal to 1 volt per ampere.

Rearranging the above equation gives $V = IR$ or $I = \dfrac{V}{R}$.

Worked example

The current through a component is 2.0 mA when the p.d. across it is 12 V. Calculate:

a its resistance at this current,

b the p.d. across the component when the current is 50 μA, assuming its resistance is unchanged.

Solution

a $R = \dfrac{V}{I} = \dfrac{12}{2.0 \times 10^{-3}} = 6000\ \Omega$

b $V = IR = 50 \times 10^{-6} \times 6000 = 0.30\ \text{V}$

Measurement of resistance

A **resistor** is a component designed to have a certain resistance which is the same regardless of the current through it. The resistance of a resistor can be measured using the circuit shown in Figure 6.8.

- The **ammeter**, A, is used to measure the current through the resistor. The ammeter must be in **series** with the resistor so the same current passes through both the resistor and the ammeter.

- The **voltmeter**, V, is used to measure the p.d. across the resistor. The voltmeter must be in **parallel** with the resistor so that they have the same p.d. Also, **no current** should pass through the voltmeter, otherwise the ammeter will not record the exact current through the resistor. In theory, the voltmeter should have **infinite resistance**. In practice, a voltmeter with a sufficiently high resistance would be satisfactory.

- The **variable resistor** is used to adjust the current and p.d. as necessary. To investigate the variation of current with p.d., the variable resistor is adjusted in steps. At each step, the current and p.d. are recorded from the ammeter and voltmeter respectively. The measurements can then be plotted on a graph of p.d. against current, as shown in Figure 6.9.

Ohm's Law

The graph for a resistor is a straight line through the origin. The resistance is the same, regardless of the current. The resistance is equal to the gradient of the graph, because the gradient is constant and at any point is equal to p.d./current.

Table 6.1 *Reminder about prefixes*

Prefix	Symbol	Value
nano	n	10^{-9}
micro	μ	10^{-6}
milli	m	10^{-3}
kilo	k	10^{+3}
mega	M	10^{+6}
giga	G	10^{+9}

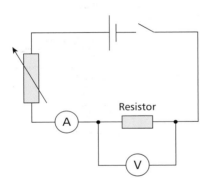

Fig 6.8 *Measuring resistance*

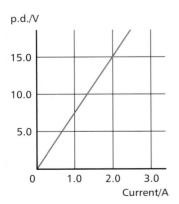

Fig 6.9 *Potential difference v. current for a resistor*

The discovery that the potential difference across a metal wire is proportional to the current through it was made by Georg Ohm in 1826 and is known as Ohm's Law.

> **Ohm's Law states that the potential difference across a metallic conductor is proportional to the current through it, provided the physical conditions do not change.**

Notes

Ohm's Law is equivalent to the statement that the resistance of a metallic conductor under **constant physical conditions** (e.g. temperature) is constant. For an **ohmic conductor**, $V = IR$ where R is constant. A resistor is a component designed to have a certain resistance.

6.3.2 Resistivity

For a conductor of length L and uniform cross-sectional area A, as shown in Figure 6.10, its resistance R is:

- proportional to L,
- inversely proportional to A.

Hence $R = \dfrac{\rho L}{A}$, where ρ is a constant for that material known as its **resistivity**.

Rearranging this equation gives the following equation, which can be used to calculate the resisitivity of a sample of material of length L and uniform cross-sectional area A:

$$\text{Resistivity, } \rho = \frac{R A}{L}$$

Notes

- The unit of resisitivity is the **ohm metre (Ω m)**.

- For a conductor with a circular cross-section of diameter d:

$$A = \frac{\pi d^2}{4} \left(= \pi r^2 \text{ where radius } r = \frac{d}{2}\right)$$

- **Conductivity,** $\sigma = \dfrac{1}{\text{resistivity}}$. The unit of conductivity is **siemens per metre (S m^{-1})**.

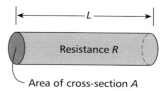

Fig 6.10 Resistivity

Table 6.2 Resistivity values of different materials at room temperature

Material	Resistivity/Ω m
Copper	1.7×10^{-8}
Constantan	5.0×10^{-7}
Carbon	3×10^{-5}
Silicon	2300.00
PVC	about 10^{14}

QUESTIONS

1 a Complete the table by calculating the missing value for each resistor:

	1	2	3	4	5
Current		2.0 A	0.45 A		5.0 mA
Potential difference	12.0 V		5.0 V	0.80 V	50 kV
Resistance		22 Ω	40 kΩ		20 MΩ

b Use Figure 6.9 to find the resistance of the resistor that gave the results shown.

2 Calculate the resistance of a uniform wire of diameter 0.32 mm and length 5.0 m. The resistivity of the material $= 5.0 \times 10^{-7}\,\Omega$ m.

3 Calculate the resistance of a rectangular strip of copper of length 0.08 m, thickness 15 mm and width 0.80 mm. The resistivity of copper $= 1.7 \times 10^{-8}\,\Omega$ m.

4 A wire of uniform diameter 0.28 mm and length 1.50 m has a resistance of 45 Ω. Calculate:

a its resistivity,

b the length of this wire that has a resistance of 1.0 Ω.

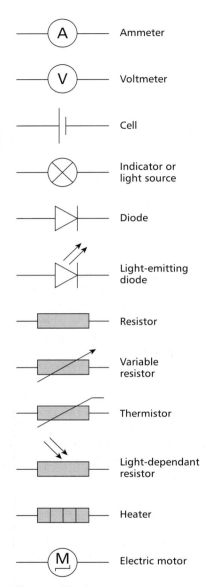

Fig 6.11 *Circuit components*

6.4.1 Circuit diagrams

Each type of component has its own **symbol**, which is used to represent the component in a **circuit diagram**. You need to recognise the symbols for different types of components to make progress – just like a motorist needs to know what different road signs mean. Note that, on a circuit diagram, the direction of the current is always shown from + **to** − round the circuit.

The function of each of the components shown in Figure 6.11 is given in the list below:

- An **ammeter** measures the current flowing through the circuit.
- A **voltmeter** measures the p.d.
- A **cell** is a source of electrical energy. Note that **a battery** is a combination of cells.
- The symbol for an **indicator** or any **light source** (including a filament lamp), except a light-emitting diode, is the same.
- A **diode** allows current in one direction only. A **light-emitting diode** (or LED) emits light when it conducts. The direction which the diode conducts in is referred to as its 'forward' direction. The opposite direction is referred to as its 'reverse' direction.
- A **resistor** is a component designed to have a certain resistance.
- A **variable resistor** is used to control the current in a circuit.
- The resistance of a **thermistor** decreases with increase of temperature, if the thermistor is an intrinsic semiconductor such as silicon. Such a thermistor is said to have a negative **temperature coefficient**. See section 6.4.3.
- The resistance of a **light-dependent resistor** decreases with increase of light intensity.
- A **heater** is designed to transform electrical energy to heat.
- An **electric motor** is designed to transform electrical energy into kinetic energy.

6.4.2 Investigating the characteristics of different components

To measure the variation of current with p.d. for a component, use either:

- a potential divider to vary the p.d. from zero (Fig 6.12a), or
- a variable resistor to vary the current to a minimum (Fig 6.12b).

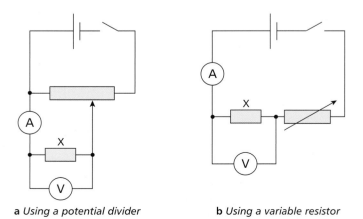

a *Using a potential divider*   **b** *Using a variable resistor*

Fig 6.12 *Investigating component characteristics*

The advantage of using a potential divider in comparison with a variable resistor is that the current through the component and the p.d. across it can be reduced to zero. However, in the variable resistor circuit, when the resistance of the variable resistor is at a maximum, the current cannot be reduced any further unless a further resistor (not shown) is connected in series with the variable resistor.

The measurements for each type of component may be plotted on a graph of current (on the *y*-axis) against p.d. (on the *x*-axis). Typical graphs for a wire, a filament lamp, and a thermistor are shown in Figure 6.13. Note that the measurements are the same, regardless of which way the current passes through each of these components.

- A **wire** gives a straight line, with a constant gradient equal to the 1/resistance *R* of the wire. Any resistor at constant temperature would give a straight line.
- A **filament bulb** gives a curve with a decreasing gradient, because its resistance increases as it becomes hotter.
- A **thermistor** at constant temperature gives a straight line. The higher the temperature, the greater the gradient of the line as the resistance falls with increase of temperature. The same result is obtained for a light-dependent resistor (LDR) in respect of light intensity.

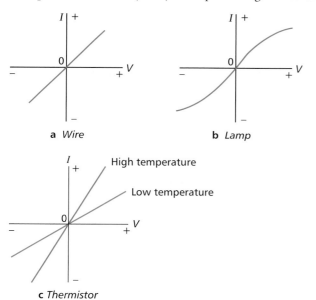

a *Wire* **b** *Lamp*

c *Thermistor*

Fig 6.13 *Current v. potential difference for various components*

The diode

To investigate the characteristics of the **diode**, one set of measurements needs to be made with the diode in its 'forward direction' (i.e. forward biased) and another set with it in its 'reverse direction' (i.e. reverse biased). The current is very small when the diode is reverse biased and can only be measured using a milliammeter.

Typical results for a silicon diode are shown in Figure 6.14. A silicon diode conducts easily in its 'forward' direction above a p.d. of about 0.6 V, and hardly at all below 0.6 V or in the opposite direction.

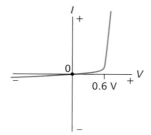

Fig 6.14 *Current v. potential difference for a silicon diode*

6.4.3 Resistance and temperature

- **The resistance of a metal** increases with increase of temperature. This is because the positive ions in the conductor vibrate more when its temperature is increased. The charge carriers (i.e. conduction electrons) therefore cannot pass through the metal as easily when a p.d. is applied across the conductor. A metal is said to have a **positive temperature coefficient** because its resistance increases with increase of temperature.
- **The resistance of an intrinsic semiconductor** decreases with increase of temperature. This is because the number of charge carriers (i.e. conduction electrons) increases when the temperature is increased. A thermistor made from an intrinsic semiconductor therefore has a **negative temperature coefficient**.

QUESTIONS

1. A filament bulb is labelled '3.0 V, 0.75 W'.
   a. Calculate its current and its resistance at 3.0 V.
   b. State and explain what would happen to the filament bulb if the current was increased from the value in part **a**.

2. A certain thermistor has a resistance of 50 000 Ω at 20 °C and a resistance of 4000 Ω at 60 °C.
   It is connected in series with an ammeter and a 1.5 V cell. Calculate the ammeter reading when the thermistor is:
   a. at 20 °C   b. at 60 °C

3. A silicon diode is connected in series with a cell and a torch bulb:
   a. Sketch the circuit diagram showing the diode in its 'forward' direction.
   b. Explain why the torch bulb would not light if the polarity of the cell was reversed in the circuit.

4. The resistance of a certain metal wire increased from 25.3 Ω at 0 °C to 35.5 Ω at 100 °C. Assuming the resistance over this range varies linearly with temperature, calculate:
   a. the resistance at 50 °C,
   b. the temperature when the resistance is 30.0 Ω.

1 A rechargeable battery is capable of delivering a current of 0.25 A for 1800 s at a potential difference of 1.5 V. Calculate:

   a the total charge the battery is capable of delivering without being recharged,

   b the energy transferred from the battery as a result of delivering the charge calculated in part **a**.

2 A torch bulb is marked with the label '0.5 W, 1.5 V'. Calculate:

   a the current through the torch bulb when it operates at normal brightness,

   b the energy transferred to the torch bulb in 5 min when it operates at normal brightness.

3 a With the aid of a circuit diagram, describe how you would investigate the variation of p.d. with current for a 12 V filament light bulb.

   b (i) Sketch a graph to show how the p.d. varies with current for a filament light bulb.

      (ii) Describe how the resistance of a filament light bulb changes with brightness.

      (iii) Explain why the resistance of a filament light bulb changes in the way described in part (ii).

4 A 3.0 V battery, a diode, a switch, an ammeter and a 200 Ω resistor are connected in series with each other.

   a Sketch the circuit diagram for this arrangement with the diode in the forward direction.

   b When the switch is closed, the ammeter records a current of 0.012 A. Calculate:

      (i) the p.d. across the resistor,

      (ii) the p.d. across the diode.

   c The 200 Ω resistor was replaced by a 50 Ω resistor. Calculate the ammeter reading when the switch is closed.

5 a With the aid of a suitable circuit diagram, describe how you would measure the resistivity of a uniform wire of known cross-sectional area.

   b A wire of length 1.20 m and area of cross-section 6.0×10^{-7} m^2 has a resistance of 1.1 Ω.

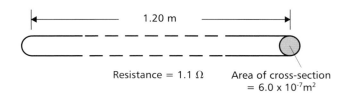

Resistance = 1.1 Ω   Area of cross-section = 6.0×10^{-7} m^2

   Calculate:
   (i) the resistivity of the material of this wire,
   (ii) the resistance of a wire consisting of this material of length 2.40 m and area of cross-section 3.0×10^{-7} m^2.

6 A thermistor, a 100 Ω resistor, a 9.0 V battery and a 0–100 mA ammeter are connected in series:

   a Sketch the circuit diagram for this arrangement.

   b (i) The ammeter reads 50 mA. State and explain how the ammeter reading would change if the temperature of the thermistor, which has a negative temperature coefficient, was raised.

      (ii) What change would need to be made to the circuit to reduce the current to 20 mA without changing the temperature of the thermistor?

7 A copper wire in a domestic electric circuit has a diameter of 1.1 mm and a length of 12 m.

   a Calculate:

      (i) the resistance of this wire,

      (ii) the p.d. across the ends of the wire when the current through it is 13 A,

      (iii) the power dissipated by the wire due to its resistance when a current of 13 A passes through it.

   b Electrical safety regulations require that the p.d. across the ends of the wire should not exceed 6.0 V. Calculate the maximum safe current through the wire. The resistivity of copper = 1.7×10^{-8} Ω m

8 a Calculate the resistance of a 230 V, 100 W filament light bulb.

   b The filament of the light bulb in part **a** consists of a coiled tungsten wire of diameter 0.05 mm. Calculate the total length of the wire of this filament. The resistivity of tungsten = 5.6×10^{-8} Ω m

9 A light-dependent resistor has a resistance of 650 Ω in darkness. It is connected in series with a 4.5 V battery, a resistor, a milliammeter and a switch, as shown below.

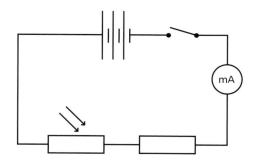

a The milliammeter reads 5.0 mA when the switch is closed and the LDR is in darkness. Calculate:

(i) the p.d. across the LDR,

(ii) the p.d. across the resistor,

(iii) the resistance of the resistor.

b Describe how, and explain why, the milliammeter reading changes when the LDR is exposed to light.

10 The following measurements were made in an investigation to measure the resistivity of the material of a certain wire:

P.d. across the wire/V	0.00	2.0	4.0	6.0	8.0	10.0
Current through the wire/V	0.00	0.15	0.31	0.44	0.62	0.74

Length of wire = 1.60 m; Diameter of wire = 0.28 mm

a Plot a graph of the p.d. against the current.

b Use the graph to calculate the resistivity of the material of the wire.

DC Circuits

Circuit rules

7.1.1 Currrent rules

Kirchhoff's First Law

> **At any junction in a circuit, the total current leaving the junction is equal to the total current entering the junction.**

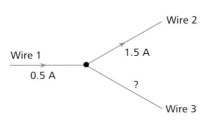

Fig 7.1 *Kirchhoff's First Law*

For example, Figure 7.1 shows a junction of three wires where the current in two of the wires (Wire 1 and Wire 2) is given. The current in Wire 3 must be 1.0 A into the junction, because the total current into the junction (1.0 A along Wire 3 + 0.5 A along Wire 1) is the same as the total current out of the junction (1.5 A along Wire 2).

Kirchhoff's First Law follows from the **conservation of charge**, as charge flow in and charge flow out of a junction are always equal. The current along a wire is the charge flow per second. In Figure 7.1, the charge entering the junction each second is 0.5 C along Wire 1 and 1.0 C along Wire 3. The charge leaving the junction each second must therefore be 1.5 C, as the junction does not retain charge.

Components in series

- **The current entering a component is the same as the current leaving the component.** In other words, components do not use up current. At any instant, the charge entering a component each second is equal to the charge leaving it because the same number of charge carriers enter and leave the component each second. In Figure 7.2, A_1 and A_2 show the same reading because they are measuring the same current.

- **The current passing through two or more components in series is the same through each component.** This is because each charge carrier passes through every component and the same number of charge carriers pass through each component each second. At any instant, charge flows at the same rate through each component.

7.1.2 Potential difference rules

Energy and potential difference

> **The potential difference, or voltage, between any two points in a circuit is defined as the energy transfer per coulomb of charge that flows from one point to the other.**

If the charge carriers lose energy, the potential difference is a potential drop. If the charge carriers gain energy, which happens when they pass through a battery or cell, the potential difference is a potential rise equal to the e.m.f. of the battery or cell. The rules for potential differences are listed below with an explanation of each rule in energy terms.

- **For two or more components in series, the total p.d. across all the components is equal to the sum of the potential differences across each component.**
  Figure 7.3 shows a circuit consisting of a battery and three resistors in series. The p.d. across the battery terminals is equal to the sum of the potential

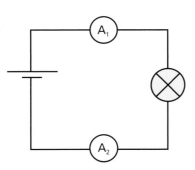

Fig 7.2 *Components in series*

differences across the three resistors. This is because each coulomb of charge from the battery delivers energy to each resistor as it flows round the circuit. The p.d. across each resistor is the energy delivered per coulomb of charge to that resistor. So the sum of the potential differences across the three resistors is the **total energy** delivered to the resistors per coulomb of charge that passes through them, which is the p.d. across the battery terminals.

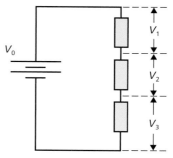

Fig 7.3 *Adding potential differences*

- **The p.d. across components in parallel is the same.**

  In Figure 7.4, charge carriers can pass through either of the two resistors in parallel. The same amount of energy is delivered by a charge carrier, regardless of which of the two resistors it passes through.

  If the variable resistor is adjusted so the p.d. across it is 4 V, and if the battery p.d. is 12 V, the p.d. across the two resistors in parallel is 8 V (12 V − 4 V). This is because each coulomb of charge leaves the battery with 12 J of electrical energy, and uses 4 J on passing through the variable resistor. Therefore, each coulomb of charge has 8 J of electrical energy to deliver to either of the two parallel resistors.

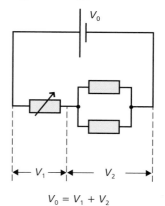

$$V_0 = V_1 + V_2$$

Fig 7.4 *Components in parallel*

- **For any complete loop of a circuit, the sum of the e.m.f.s round the loop is equal to the sum of the potential drops round the loop.**

  This statement is known as **Kirchhoff's Second Law**. It follows from the fact that the total e.m.f. in a loop is the total electrical energy per coulomb produced in the loop, and the sum of the potential drops is the electrical energy per coulomb delivered round the loop. The above statement follows, therefore, from the **conservation of energy**.

  For example, Figure 7.5 shows a 9 V battery connected to a 6 V light bulb in series with a variable resistor. If the variable resistor is adjusted so that the p.d. across the light bulb is 6 V, the p.d. across the variable resistor must be 3 V (9 V − 6 V). The only source of electrical energy in the circuit is the battery, so the sum of the e.m.f.s in the circuit is 9 V. This is equal to the sum of the p.ds round the circuit (3 V across the variable resistor + 6 V across the light bulb). In other words, the battery forces charge round the circuit. Every coulomb of charge leaves the battery with 9 J of electrical energy and supplies 3 J to the variable resistor and 6 J to the light bulb.

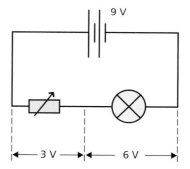

Fig 7.5 *Applying Kirchhoff's Second Law*

QUESTIONS

1 A battery, which has an e.m.f. of 6 V and negligible internal resistance, is connected to a 6 V, 6 W light bulb in parallel with a 6 V, 24 W light bulb, as shown in Figure 7.6.

   Calculate:

   a the current through each light bulb,

   b the current from the battery,

   c the power supplied by the battery.

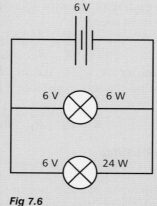

Fig 7.6

2 A 4.5 V battery is connected in series with a variable resistor and a 2.5 V, 0.5 W torch bulb.

   a Sketch the circuit diagram for this circuit.

   b The variable resistor is adjusted so that the p.d. across the torch bulb is 2.5 V. Calculate:
   (i) the p.d. across the variable resistor,
   (ii) the current through the torch bulb.

3 A 6.0 V battery is connected in series with an ammeter, a 20 Ω resistor and an unknown resistor R.

   a Sketch the circuit diagram.

   b The ammeter reads 0.20 A. Calculate:
   (i) the p.d. across the 20 Ω resistor,
   (ii) the p.d. across R,
   (iii) the resistance of R.

4 In question **3**, when the unknown resistor is replaced with a torch bulb, the ammeter reads 0.12 A. Calculate:

   a the p.d. across the torch bulb,

   b the resistance of the torch bulb.

7.2 More about resistance

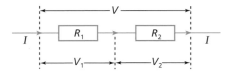

Fig 7.7 Resistors in series

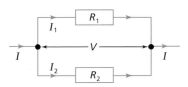

Fig 7.8 Resistors in parallel

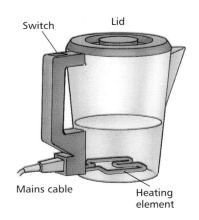

Switch | Lid | Mains cable | Heating element

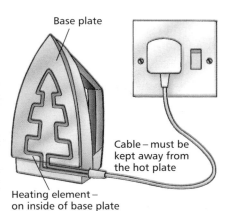

Base plate | Cable – must be kept away from the hot plate | Heating element – on inside of base plate

Fig 7.9 Heating elements

7.2.1 Resistor combination rules

Resistors in series

Resistors in series pass the same current. The total p.d. is equal to the sum of the individual potential differences.

- For two resistors R_1 and R_2 in series, as in Figure 7.7, when current I passes through the resistors:

  p.d. across R_1, $V_1 = IR_1$
  p.d. across R_2, $V_2 = IR_2$

  The total p.d. across the two resistors, $V = V_1 + V_2 = IR_1 + IR_2$

  therefore, the total resistance $R = \dfrac{V}{I} = \dfrac{IR_1 + IR_2}{I} = R_1 + R_2$

- For two or more resistors R_1, R_2, R_3, etc. in series, the theory can easily be extended to show that the **total resistance is equal to the sum of the individual resistances**:

$$R = R_1 + R_2 + R_3 + ...$$

Resistors in parallel

Resistors in parallel have the same p.d. The current through a parallel combination of resistors is equal to the sum of the individual currents.

- For two resistors R_1 and R_2 in parallel, as in Figure 7.8, when the p.d. across the combination is V,

  the current through resistor R_1, $I_1 = \dfrac{V}{R_1}$

  the current through resistor R_2, $I_2 = \dfrac{V}{R_2}$

  The total current through the combination, $I = I_1 + I_2 = \dfrac{V}{R_1} + \dfrac{V}{R_2}$

  Since the total resistance $R = \dfrac{V}{I}$, then the total current $I = \dfrac{V}{R}$

  therefore $\dfrac{V}{R} = \dfrac{V}{R_1} + \dfrac{V}{R_2}$

Cancelling V from each term gives the following equation, which is used to calculate the total resistance R:

$$\dfrac{1}{R} = \dfrac{1}{R_1} + \dfrac{1}{R_2}$$

- For two or more resistors R_1, R_2, R_3, etc. in parallel, the theory can easily be extended to show that the total resistance R is given by:

$$\dfrac{1}{R} = \dfrac{1}{R_1} + \dfrac{1}{R_2} + \dfrac{1}{R_3} + ...$$

7.2.2 Resistance heating

The heating effect of an electric current in any component is due to the resistance of the component. As explained in section 6.2, the charge carriers repeatedly collide with the positive ions of the conducting material. There is a **net transfer of energy** from the charge carriers to the positive ions as a result of these collisions. After a charge carrier loses kinetic energy in such a collision, the force due to the p.d. across the material accelerates it until it collides with another positive ion.

For a component of resistance R, when current I passes through it:

p.d. across the component, $V = IR$

Therefore the power supplied to the component, $P = IV = I^2R \left(= \dfrac{V^2}{R} \right)$

Hence the **energy per second** transferred to the component as thermal energy $= I^2R$

- If the component is at constant temperature, heat transfer to the surroundings takes place at the same rate. Therefore:

Rate of heat transfer $= I^2R$

- If the component heats up, its temperature rise depends on the power supplied to it (i.e. I^2R), the rate of heat transfer to the surroundings and the heat capacity of the component.

- The energy transfer per second to the component (i.e. the power supplied to it) does not depend on the direction of the current. For example, the heating effect of an alternating current at a given instant depends **only on the magnitude of the current** not on the direction of the current.

QUESTIONS

1 Calculate the total resistance of each of the resistor combinations in Figure 7.10.

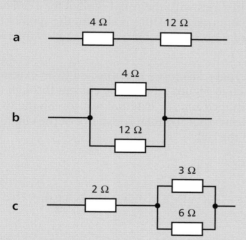

Fig 7.10

2 A 3 Ω resistor and a 6 Ω resistor are connected in parallel with each other. The parallel combination is connected in series with a 6 V battery of negligible internal resistance and a 4 Ω resistor, as shown in Figure 7.11.

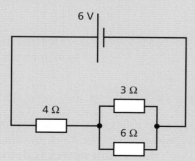

Fig 7.11

Calculate:

a the combined resistance of the 3 Ω resistor and the 6 Ω resistor in parallel,

b the total resistance of the circuit,

c the battery current,

d the power supplied to the 4 Ω resistor.

3 A 2 Ω resistor and a 4 Ω resistor are connected in series with each other. The series combination is connected in parallel with a 9 Ω resistor and a 3 V battery of negligible internal resistance, as shown in Figure 7.12.

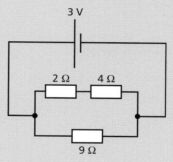

Fig 7.12

Calculate:

a the total resistance of the circuit,

b the battery current,

c the power supplied to each resistor,

d the power supplied by the battery.

4 Calculate:

a the power supplied to a 10 Ω resistor when the p.d. across it is 12 V,

b the resistance of a heating element designed to operate at 60 W and 12 V.

E.m.f. and internal resistance

7.3.1 Internal resistance

The **internal resistance** of a source of electricity is due to opposition to the flow of charge through the source. This causes electrical energy produced by the source to be dissipated inside the source when charge flows through it.

The **electromotive force (e.m.f.)** of the source is the electrical energy per unit charge produced by the source.

The **p.d. across the terminals** of the source is the electrical energy per unit charge that can be delivered by the source when it is in a circuit.

The **terminal p.d.** is less than the e.m.f. whenever current passes through the source. The difference is due to the internal resistance of the source.

> **The internal resistance of a source is the loss of potential difference per unit current in the source, when current passes through the source.**

In circuit diagrams, the internal resistance of a source may be shown as a resistor (labelled 'internal resistance') in series with the usual symbol for a cell or battery, as in Figure 7.13. If no internal resistance is shown, the symbol for the cell or battery should be labelled with symbols for its e.m.f. and its internal resistance.

When a cell of e.m.f. ε and internal resistance r is connected to an external resistor of resistance R, as shown in Figure 7.14, all the current through the cell passes through its internal resistance and the external resistor. So the two resistors are in series, which means that the total resistance of the circuit is $r + R$. Therefore, the current through the cell:

$$I = \frac{\varepsilon}{R + r}$$

In other words, the cell e.m.f., $\varepsilon = I(R + r) = IR + Ir$ = the cell p.d. + the 'lost' p.d.

$$\boldsymbol{\varepsilon = IR + Ir}$$

The 'lost' p.d. inside the cell (i.e. the p.d. across the internal resistance of the cell) is equal to the difference between the cell e.m.f. and the p.d. across its terminals. In energy terms, the 'lost' p.d. is the energy per coulomb dissipated or wasted inside the cell due to its internal resistance.

Power

Multiplying each term of the above equation by the cell current I gives:

Power supplied by the cell, $I\varepsilon = I^2R + I^2r$

In other words,

Power supplied by the cell = power delivered to R + power wasted in the cell due to its internal resistance

The power delivered to R $= I^2R = \dfrac{\varepsilon^2 R}{(R + r)^2}$ since $I = \dfrac{\varepsilon}{R + r}$

Figure 7.15 shows how the power delivered to R varies with the value of R. It can be shown that the peak of this power curve is at $R = r$. In other words, when a source delivers power to a 'load', **maximum power is delivered to the load when the load resistance is equal to the internal resistance of the source.** The load is then said to be 'matched' to the source.

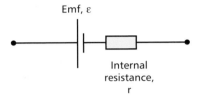

Emf, ε

Internal resistance, r

Fig 7.13 Internal resistance

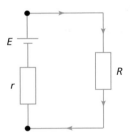

E

r

R

Fig 7.14 E.m.f. and internal resistance

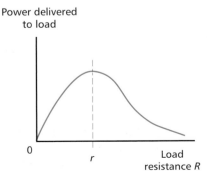

Power delivered to load

0   r   Load resistance R

Fig 7.15 Power delivered to a load v. load resistance

7.3.2 Measurement of internal resistance

The potential difference across the terminals of a cell when the cell is in a circuit can be measured by connecting a high-resistance voltmeter directly across the terminals of the cell. Figure 7.16 shows how the cell p.d. can be measured for different values of current. The current is changed by adjusting the variable resistor. The lamp limits the maximum current that can pass through the cell. The ammeter is used to measure the cell current.

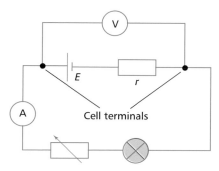

Fig 7.16 Measuring internal resistance

Graph of p.d. v. current

The measurements of cell p.d. and current for a given cell may be plotted on a graph, as shown in Figure 7.17.

The cell p.d. decreases as the current increases. This is because the 'lost' p.d. increases as the current increases.

- **The cell p.d. is equal to the cell e.m.f. at zero current**. This is because the 'lost' p.d. is zero at zero current.

- **The graph is a straight line with a negative gradient**. This can be seen by rearranging the equation:

$$\varepsilon = IR + Ir \text{ to become } IR = \varepsilon - Ir$$

Because IR represents the cell p.d. V, then $V = \varepsilon - Ir$. By comparison with the standard equation for a straight line $y = mx + c$, a graph of V on the y-axis against I on the

x-axis gives a straight line with a gradient $-r$ and a y-intercept ε. See p.157.

Figure 7.17 shows the gradient triangle ABC, in which AB represents the lost p.d. and BC represents the current.

So the gradient $\dfrac{AB}{BC} = \dfrac{\text{lost voltage}}{\text{current}} =$ internal resistance r.

Note The internal resistance and the e.m.f. of a cell can be calculated, if the cell p.d. is measured for two different values of current. A pair of simultaneous equations can therefore be written, as follows:

- For current I_1, the cell p.d. $V_1 = \varepsilon - I_1 r$
- For current I_2, the cell p.d. $V_2 = \varepsilon - I_2 r$

Subtracting the first equation from the second gives:

$$V_1 - V_2 = (\varepsilon - I_1 r) - (\varepsilon - I_2 r) = I_2 r - I_1 r = (I_2 - I_1)r$$

Therefore, $r = \dfrac{V_1 - V_2}{(I_2 - I_1)}$

So r can be calculated from the above equation and then substituted into either equation for the cell p.d. to enable ε to be calculated.

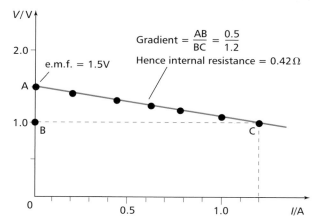

Fig 7.17 A graph of cell p.d. v. current

QUESTIONS

1 A battery of e.m.f. 12 V and internal resistance 1.5 Ω was connected to a 4.5 Ω resistor. Calculate:

   a the total resistance of the circuit,

   b the current through the battery,

   c the lost p.d.,

   d the p.d. across the cell terminals.

2 A cell of e.m.f. 1.5 V and internal resistance 0.5 Ω is connected to a 2.5 Ω resistor. Calculate:

   a the current,

   b the terminal p.d.,

   c the power delivered to the 2.5 Ω resistor,

   d the power wasted in the cell.

3 The p.d. across the terminals of a cell was 1.1 V when the current from the cell was 0.20 A and 1.3 V when the current was 0.10 A. Calculate:

   a the internal resistance of the cell,

   b the cell's e.m.f.

4 A battery of unknown e.m.f., ε, and internal resistance, r, is connected in series with an ammeter and a resistance box, R. The current was 2.0 A when $R = 4.0$ Ω, and 1.5 A when $R = 6.0$ Ω. Calculate ε and r.

More circuit calculations

7.4.1 Circuits with a single cell and one or more resistors

Here are some rules:

- Sketch the **circuit diagram** if it is not drawn.
- To calculate the **current** passing through the cell, calculate the total circuit resistance using the resistor combination rules. Don't forget to add on the internal resistance of the cell, if that is given:

$$\text{Cell current} = \frac{\text{cell e.m.f.}}{\text{total circuit resistance}}$$

- To work out the **current and p.d.** for each resistor, start with the **resistors in series** with the cell which pass the same current as the cell current:

  P.d. across each resistor in series with cell = current $\times$ the resistance of each resistor

- To work out the current through **parallel resistors**, work out the combined resistance (product/sum) and multiply by the cell current to give the p.d. across each resistor:

$$\text{Current through each resistor} = \frac{\text{p.d. across parallel combination}}{\text{resistor's resistance}}$$

7.4.2 Circuits with two or more cells in series

The same rules as above apply, except the current through the cells is calculated by dividing the **overall (i.e. net) e.m.f.** by the total resistance.

- If the cells are connected in the **same** direction in the circuit (Fig 7.18a), the net e.m.f. is the **sum** of the individual e.m.fs. For example, in Figure 7.18a the net e.m.f. is 3.5 V.
- If the cells are connected in **opposite** directions to each other in the circuit (Fig 7.18b), the net e.m.f. is the **difference** between the e.m.fs in each direction. For example, in Figure 7.18b, the net e.m.f. is 0.5 V in the direction of the 2.0 V cell.
- The **total internal resistance** is the **sum** of the individual internal resistances. This is because the cells, and therefore the internal resistances, are in series.

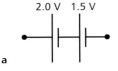

a

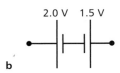

b

Fig 7.18 Cells in series

Worked example

A battery of e.m.f. 3.0 V (and internal resistance 2.0 Ω) and a battery of e.m.f. 2.0 V (and internal resistance 1.0 Ω) are connected in series with each other and with a 7.0 Ω resistor (Fig 7.19). Calculate the p.d. across the 7.0 Ω resistor.

Solution

Net e.m.f. of the two batteries = 3.0 − 2.0 = 1.0 V in the direction of the 3.0 V battery

Total circuit resistance = 1.0 Ω + 2.0 Ω + 7.0 Ω = 10.0 Ω

Therefore $\qquad \text{Battery current} = \dfrac{\text{net e.m.f.}}{\text{total circuit resistance}}$

$$= \frac{1.0 \text{ V}}{10.0 \text{ Ω}} = 0.10 \text{ A}$$

p.d. across the 7.0 Ω resistor = current $\times$ resistance

$$= 0.10 \text{ A} \times 7.0 \text{ Ω} = 7.0 \text{ V}$$

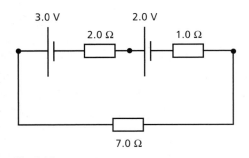

Fig 7.19

7.4.3 Diodes in circuits

Assume that a semiconductor diode has:

- **zero resistance in the forward direction** when the p.d. across it is 0.6 V or greater,
- **infinite resistance in the reverse direction** or at p.ds less than 0.6 V in the forward direction.

Therefore, in a circuit with one or more diodes as above:

- a p.d. of 0.6 V exists across a diode that is forward-biased and passing a current,
- a diode that is reverse-biased has infinite resistance.

For example, suppose a diode is connected in its forward direction in series with a 1.5 V cell and a 1.5 kΩ resistor, as in Figure 7.20.

The p.d. across the diode is 0.6 V, because it is forward-biased. Therefore, the p.d. across the resistor is 0.9 V (= 1.5 V − 0.6 V). The current through the resistor is therefore $6.0 \times 10^{-4} \text{ A} \left(= \dfrac{0.9 \text{ V}}{1500 \text{ }\Omega} \right)$.

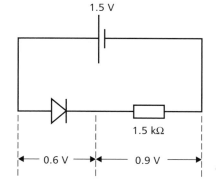

Fig 7.20 *Using a diode*

QUESTIONS

1 A cell of e.m.f. 3.0 V, and negligible internal resistance, is connected to a 4.0 Ω resistor in series with a parallel combination of a 24.0 Ω resistor and a 12.0 Ω resistor (Fig 7.21).

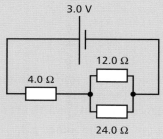

Fig 7.21

Calculate:

a the total resistance of the circuit,

b the cell current,

c the current and p.d. for each resistor.

2 A battery of e.m.f. 12.0 V, with an internal resistance of 3.0 Ω, is connected in series with a 15.0 Ω resistor and a battery of e.m.f. 9.0 V, which has an internal resistance of 2.0 Ω (Fig 7.22). Calculate:

a the total resistance of the circuit,

b the cell current,

c the current and p.d. across the 15 Ω resistor.

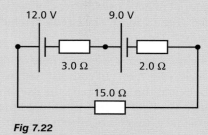

Fig 7.22

3 a Two 8 Ω resistors and a battery of e.m.f. 12.0 V, and internal resistance 8 Ω, are connected in series with each other. Sketch the circuit diagram and calculate:
  (i) the power delivered to each external resistor,
  (ii) the power wasted due to internal resistance.

  b The two 8 Ω resistors in part a are reconnected in parallel with each other and then connected to the same battery. Sketch the circuit diagram and calculate:
  (i) the power delivered to each external resistor,
  (ii) the power wasted due to internal resistance.

4 a For the circuit shown in Figure 7.23, calculate the p.d. and current for each resistor and diode.

  b In the circuit (Fig 7.23), both diodes are reversed. Sketch the new circuit and calculate the p.d. and current for each resistor and diode for this new arrangement.

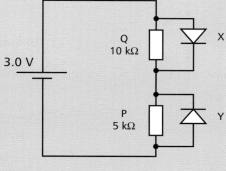

Fig 7.23

7.5.1 The theory of the potential divider

A **potential divider** consists of two or more resistances in series with each other and with a source of fixed potential difference. The potential difference of the source is divided beween the components in the circuit, as they are in series with each other. By making a suitable choice of components, a potential divider can be used:

- to supply a p.d. which is fixed at any value between zero and the source p.d.
- to supply a variable p.d.
- to supply a p.d. that varies with a physical condition, such as temperature or light intensity.

To supply a fixed p.d.

Consider two resistors R_1 and R_2, in series and connected to a source of fixed p.d. V_0, as shown in Figure 7.24.

Total resistance of the combination $= R_1 + R_2$

Therefore, current through the resistors, $I = \dfrac{\text{p.d. across the resistors}}{\text{total resistance}} = \dfrac{V_0}{R_1 + R_2}$

so the p.d. across resistor R_1, $V_1 = IR_1 = \dfrac{V_0 R_1}{(R_1 + R_2)}$

and the p.d. across resistor R_2, $V_2 = IR_2 = \dfrac{V_0 R_2}{(R_1 + R_2)}$

These two equations show that the p.d. across each resistor, as a proportion of the source p.d., is the same as the resistance of the resistor in proportion to the total resistance. In other words, if the resistances are 5 kΩ and 10 kΩ respectively:

- The p.d. across the 5 kΩ resistor is $\frac{1}{3}$ ($= \frac{5}{15}$) of the source p.d.
- The p.d. across the 10 kΩ resistor is $\frac{2}{3}$ ($= \frac{10}{15}$) of the source p.d.

Also, dividing the equation for V_1 by the equation for V_2 gives:

$$\frac{V_1}{V_2} = \frac{R_1}{R_2}$$

This equation shows that:

> **The ratio of the p.ds across each resistor is equal to the resistance ratio of the two resistors.**

To supply a variable p.d.

The source p.d. is connected to a fixed length of uniform resistance wire. A sliding contact on the wire can then be moved along the wire, as illustrated in Figure 7.25, giving a variable p.d. between the contact and one end of the wire. A uniform track of a suitable material may be used instead of resistance wire. The track may be linear or circular (Fig 7.25a,b). The circuit symbol for a variable potential divider is shown in Figure 7.25c.

A variable potential divider can be used to vary the brightness of a light bulb between zero and normal brightness (Fig 7.26). In contrast with using a variable resistor in series with the light bulb and the source p.d., the use of a potential divider enables the current through the light bulb to be reduced to zero. With a variable resistor at maximum resistance, there is a current through the light bulb.

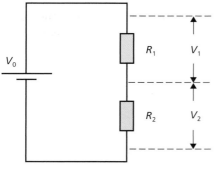

Fig 7.24 A potential divider

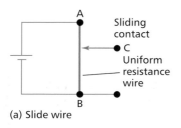

(a) Slide wire

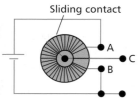

(b) Circular track

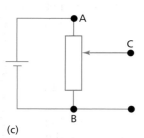

(c)

Fig 7.25 Potential dividers used to supply a variable p.d.

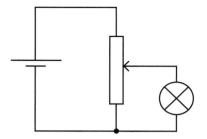

Fig 7.26 *Brightness control using a variable potential divider*

7.5.2 Sensor circuits

A **sensor circuit** produces an output p.d. which changes as a result of a change of a physical variable, such as temperature or pressure.

A temperature sensor

This consists of a potential divider made using a **thermistor** and a variable resistor (Fig 7.27).

With the temperature of the thermistor constant, the source p.d. is divided between the thermistor and the variable resistor. By adjusting the variable resistor, the p.d. across the thermistor can then be set at any desired value. When the temperature of the thermistor changes, its resistance changes so the p.d. across it changes. For example, suppose the variable resistor is adjusted so that the p.d. across the thermistor at 20 °C is exactly half the source p.d.; if the temperature of the thermistor is then raised, its resistance falls so the p.d. across it falls.

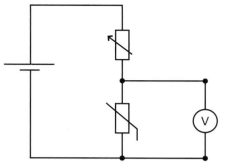

Fig 7.27 *A temperature sensor*

A light sensor

This uses a **light-dependent resistor** and a variable resistor (Fig 7.28). The p.d. across the LDR changes when the incident light intensity on the LDR changes. If the light intensity increases, the resistance of the LDR falls and the p.d. across the LDR falls.

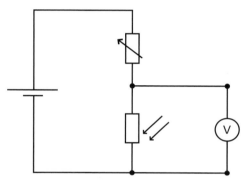

Fig 7.28 *A light sensor*

QUESTIONS

1 A potential divider consists of a 1.0 kΩ resistor in series with a 5.0 kΩ resistor, and a battery of e.m.f. 4.5 V and negligible internal resistance.

   a Sketch the circuit and calculate the p.d. across each resistor.

   b A second 5.0 kΩ resistor is connected in the above circuit in parallel with the first 5.0 kΩ resistor. Calculate the p.d. across each resistor in this new circuit.

2 A 12 V battery, of negligible internal resistance, is connected to the fixed terminals of a variable potential divider (which has a maximum resistance of 50 Ω). A 12 V light bulb is connected between the sliding contact and the negative terminal of the potential divider. Sketch the circuit diagram and describe how the brightness of the light bulb changes when the sliding contact is moved from the negative to the positive terminal of the potential divider.

3 **a** A potential divider consists of an 8.0 Ω resistor in series with a 4.0 Ω resistor and a 6.0 V battery. Calculate:
   (i) the current,
   (ii) the p.d. across each resistor.

   b In the circuit in part **a**, the 4 Ω resistor is replaced by a thermistor with a resistance of 8 Ω at 20 °C and a resistance of 4 Ω at 100 °C. Calculate the p.d. across the fixed resistor at: (i) 20 °C, (ii) 100 °C.

4 A light sensor consists of a 5.0 V cell, an LDR and a 5.0 kΩ resistor in series with each other. A voltmeter is connected in parallel with the resistor. When the LDR is in darkness, the voltmeter reads 2.2 V.

   a Calculate:
   (i) the p.d. across the LDR,
   (ii) the resistance of the LDR when the voltmeter reads 2.2 V.

   b Describe and explain how the voltmeter reading would change if the LDR was exposed to daylight.

1 Two resistors of resistances 6.0 Ω and 12.0 Ω, in parallel, are connected to a 2.0 Ω resistor and a 6.0 V battery. The diagram below shows the circuit diagram.

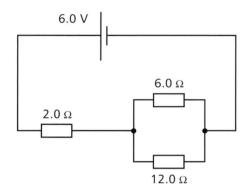

a Calculate the combined resistance of the three resistors.

b For each resistor, calculate:
  (i) the current,
  (ii) the p.d.,
  (iii) the power dissipated.

2 A light bulb is connected in series with a resistor and a battery, of negligible internal resistance, as shown below. A variable resistor is connected in parallel with the light bulb.

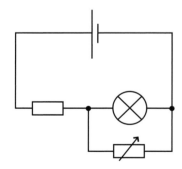

a Describe how the brightness of the light bulb changes as the resistance of the variable resistance is increased from zero.

b Explain why it is possible to reduce the bulb current to zero by adjusting the variable resistor.

3 A 15.0 Ω resistor and a 3.0 Ω resistor are connected in series with each other and with a 3.0 V battery, which has an internal resistance of 2.0 Ω, as shown in the next column.

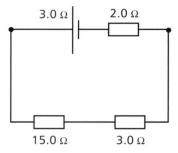

a Calculate:
  (i) the total resistance of the circuit,
  (ii) the p.d. across each of the external resistors,
  (iii) the p.d. across the battery terminals.

b Calculate the power dissipated:
  (i) in each resistor,
  (ii) in the battery due to its internal resistance.

4 The circuit diagram below is for the rear windscreen heater of a car. It consists of 4 heating elements, each of resistance 6.0 Ω, connected to a 12.0 V battery of internal resistance 2.0 Ω.

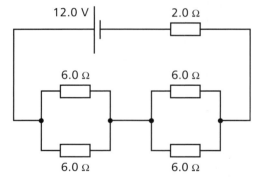

a Calculate:
  (i) the current,
  (ii) the p.d. across each heating element.

b Calculate the power dissipated:
  (i) in each heating element,
  (ii) in the battery due to its internal resistance.

c How would the operation of the heater differ if the heating elements were in parallel with each other, with the battery connected across the parallel combination?

5 A 6.0 V battery of unknown internal resistance was connected in series with a switch, a resistor of resistance 4.0 Ω and an ammeter. When the switch was closed, the ammeter read 1.0 A.

a Calculate the internal resistance of the battery.

b A second 4.0 Ω resistor was connected in parallel with the first resistor. Calculate the ammeter reading with this second resistor in the circuit.

6 A hand-held hair dryer has two 570 Ω identical heating elements, each in series with a switch. The heating elements are connected in parallel with each other and a 230 V mains electricity supply. A 12 Ω, 230 V electric fan in the hair dryer is used to blow air over the heating elements.

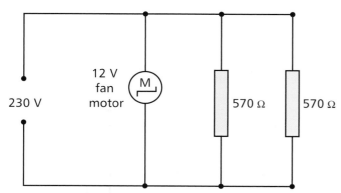

a When both heaters are being used, and the electric fan is on, calculate:
  (i) the current passing through each heating element,
  (ii) the power supplied by the electricity supply,
  (iii) the total current passing through the hair dryer.

b When one heater only is on, calculate:
  (i) the total current passing through the heater,
  (ii) the power supplied by the electricity supply.

7 The diagram shows a potential divider circuit consisting of a 10 kΩ resistor R connected in series with an n.t.c. thermistor and a 6.0 V battery. A high resistance voltmeter is connected across the thermistor.

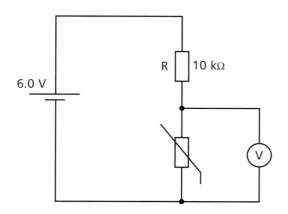

a The voltmeter reads 2.0 V when the thermistor's temperature is 20 °C. Calculate the resistance of the thermistor at this temperature.

b A second 10 kΩ resistor was connected in parallel with R. Calculate the voltmeter reading with this second resistor in the circuit when the thermistor is at the same temperature.

c Describe and explain how the voltmeter reading would change if the the temperature of the thermistor were increased.

8 The diagram shows a circuit in which two 5.0 kΩ resistors are connected in series with each other and a 5.0 V battery, of negligible internal resistance. A diode, is connected in parallel with each resistor, as shown.

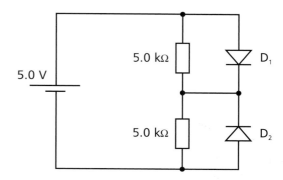

a Calculate:
  (i) the p.d.,
  (ii) the current through each resistor.

b Calculate the current through each resistor if the 5.0 V battery was replaced with a 9.0 V battery, also of negligible internal resistance.

8.1.1 Lines of force

- A **magnetic field** is a force field surrounding a magnet or current-carrying wire which acts on any other magnet or current-carrying wire placed in the field.
- The magnetic field of a bar magnet is strongest near its ends, which are referred to as '**poles**'. A bar magnet free to turn horizontally about its centre aligns itself with one end pointing North and the other pointing South. This occurs because the Earth's magnetic field attracts one end and repels the other end. The poles are referred to as **north-seeking** and **south-seeking**, according to the direction in which each pole points.
- A line of force of a magnetic field is a line along which a 'free' north pole would move in the field. Lines of force are often referred to as '**magnetic field lines**'. Note that the lines of force of a permanent magnet loop round from the north pole to the south pole of the magnet. A plotting compass points in the direction of a line of force.

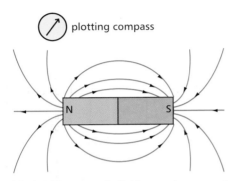

Fig 8.1 *The magnetic field near a bar magnet*

The force between two magnets

Two bar magnets placed end-to-end attract or repel, depending on whether the nearest poles are:

- **like polarity**, in which case they repel, or
- **unlike polarity**, in which case they attract.

8.1.2 Electromagnetism

A magnetic field is created round a wire whenever **a current passes** along the wire. The pattern of the magnetic field lines for a long straight wire, a solenoid and a flat coil are shown opposite. Note that the lines of force are **complete loops**. Also, the **direction of the lines of force** depends on the direction of current. If the current is reversed, the direction of the lines of force is reversed.

- **For a long straight wire**, the lines of force are circles centred on the wire in a plane perpendicular to the wire. The direction of the lines of force depends on the current direction and can be worked out using the 'corkscrew rule' as shown in Figure 8.2.
- **For a solenoid**, the lines of force pass through the solenoid along its axis and loop round outside the solenoid. The direction of the lines of force depends on the current direction and can be worked out using the solenoid rule, as shown in Figure 8.3.

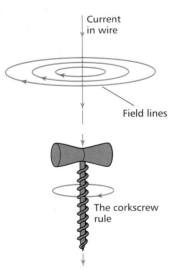

Fig 8.2 *The magnetic field near a wire*

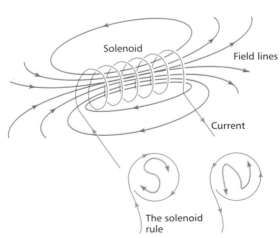

Fig 8.3 *The magnetic field of a solenoid*

- **For a flat coil**, the lines of force are lines that pass through the coil and loop round outside the coil. The direction of the lines of force can be worked out from the current direction using the solenoid rule, as explained on p. 102.

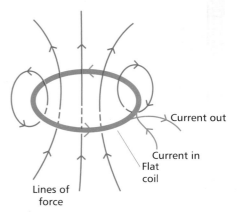

Fig 8.4 The magnetic field of a flat coil

QUESTIONS

1 A plotting compass is placed at the intersection of two perpendicular lines drawn on a sheet of paper. The plotting compass points North along one of the lines. When a bar magnet is placed along the other line near the plotting compass, as shown in Figure 8.5, the plotting compass points north-east.

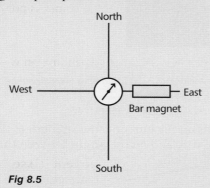

Fig 8.5

a What is the polarity of the pole of the magnet nearest the plotting compass?

b If the magnet is turned round, what direction will the plotting compass then point to?

2 An underground cable is aligned horizontally in an east-west direction. The cable carries a direct current from west to east, as shown in Figure 8.6.

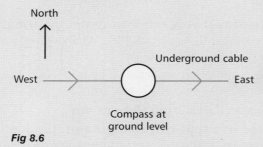

Fig 8.6

a What would be the direction of a magnetic compass directly above the wire, assuming the magnetic field due to the cable is much stronger than the Earth's magnetic field?

b How would the direction of the compass change if the current in the cable is gradually reduced to zero?

3 A plotting compass is placed at point P at the end of a solenoid. When a direct current is passed through the solenoid, the compass points into the solenoid, as shown in Figure 8.7.

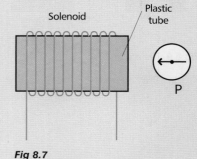

Fig 8.7

a State the direction of the current round the solenoid, as seen by an observer looking directly at the end of the solenoid at P.

b How would the direction of the plotting compass change if the current in the solenoid was reversed?

4 A student makes a model ammeter using a pair of flat coils in series with each other and a plotting compass, as shown in Figure 8.8.

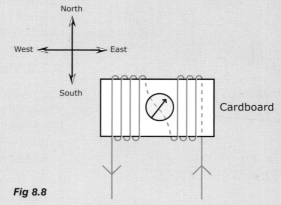

Fig 8.8

a Explain why the compass needle deflects when a direct current is passed through the coils.

b Explain why the compass needle cannot deflect more than 90° no matter how much current is passed through the coils.

The motor effect

8.2.1 The force on a current-carrying wire in a magnetic field

A current-carrying wire placed at a non-zero angle to the lines of force of an external magnetic field experiences a force due to the field. This effect is known as the **motor effect.** The force is perpendicular to the wire and to the lines of force.

The motor effect can be tested using the simple arrangement shown in Figure 8.9. The wire is placed between opposite poles of a U-shaped magnet so it is at right angles to the lines of force of the magnetic field. When a current is passed through the curve, the section of the wire in the magnetic field experiences a force that pushes it out of the field. The combined magnetic field due to the wire and the magnet is stronger on one side of the wire than on the opposite side. The wire is pushed in the direction where the combined field is weakest, as shown in Figure 8.10.

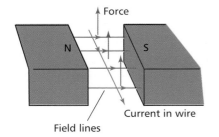

Fig 8.9 The motor effect

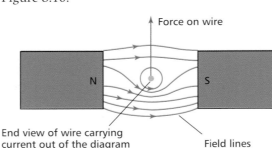

End view of wire carrying current out of the diagram   Field lines   *Fig 8.10 A field pattern*

Force factors

The magnitude of the force depends on the **current**, the strength of the **magnetic field**, the **length** of the wire and on the **angle** between the lines of force of the field and the current direction.

The force is:

* greatest when the wire is at **right angles** to the magnetic field,
* zero when the wire is **parallel** to the magnetic field.

The direction of the force is **perpendicular** to the direction of the field and to the direction of the current, as indicated by Fleming's left-hand rule, shown in Figure 8.11. If the current is reversed, or if the magnetic field is reversed, the direction of the force is reversed.

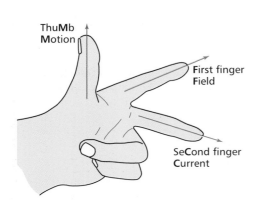

Fig 8.11 Fleming's left-hand rule

8.2.2 The electric motor

The simple electric motor consists of a coil of insulated wire, the **armature**, which spins between the poles of a U-shaped magnet. When a direct current passes round the coil:

* The wires at opposite edges of the coil are acted on by forces in opposite directions.
* The force on each edge makes the coil **spin** about its axis.

Current is supplied to the coil via a **split-ring commutator**. The direction of the current round the coil is reversed by the split-ring commutator each time the coil rotates through half a turn. This ensures the current along an edge changes direction when it moves from one pole face to the other. The result is that the force on each edge continues to turn the coil in the same direction (Fig 8.12).

The **direction of rotation** of the motor is reversed by either reversing the current **or** the direction of the magnetic field. If both the current and the

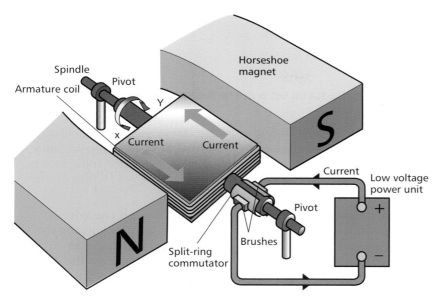

Fig 8.12 The simple electric motor

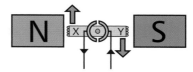

Initially, current is up side X and down side Y. Therefore the coil turns clockwise

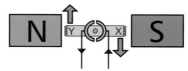

After half a turn, current is up side Y and down side X. Therefore the coil continues to turn clockwise

magnetic field are reversed, the direction of rotation is unchanged.

The **speed of rotation** depends on the **current**, the **strength** of the magnet and the **number of turns** of the coil. The strength of the magnetic field is increased if an armature with an iron core is used.

A practical electric motor

A practical electric motor has an armature with several **evenly spaced coils** wound on it. Each coil is connected to its own section of the commutator. The result is that each coil in sequence experiences a turning effect when it is connected to the voltage supply, so the motor runs smoothly.

Fig 8.13 A practical electric motor

QUESTIONS

1 A fixed vertical wire is in a horizontal magnetic field. State the direction of the force on the wire if :

a the current is upwards and the magnetic field lines are from East to West,

b the current is upwards and the magnetic field lines are from South to North,

c the current is downwards and the magnetic field lines are from East to West.

2 A fixed horizontal wire lies along a line from East to West in a magnetic field. State the direction of the magnetic field lines if the current in the wire is from East to West and the force on it is:

a vertically up,

b horizontal and due North.

3 A rectangular coil carries a current in a magnetic field:

a Explain why the coil experiences a turning effect when the plane of the coil is parallel to the magnetic field lines, as in Figure 8.14a.

b Explain why the coil experiences no turning effect when the plane of the coil is perpendicular to the magnetic field lines, as in Figure 8.14b

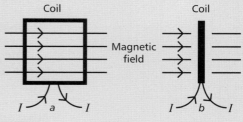

Fig 8.14

4 a What is the function of the split-ring commutator in a d.c. electric motor?

b A simple electric motor containing a permanent magnet is connected to a battery and a variable resistor. What would be the effect on the motor of:

(i) increasing the current,

(ii) reversing the current,

(iii) using an a.c. supply instead of the battery?

Magnetic flux density

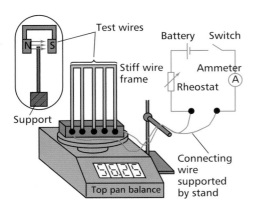

Fig 8.15 Measuring the force on a current-carrying wire in a magnetic field

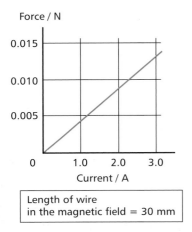

Length of wire
in the magnetic field = 30 mm

Fig 8.16 Force v. current

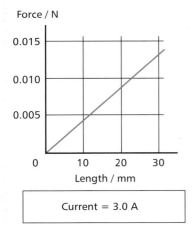

Current = 3.0 A

Fig 8.17 Force v. length

8.3.1 Investigating the force on a current-carrying wire in a magnetic field

The magnitude of the force on a current-carrying wire in a magnetic field can be investigated using the arrangement shown in Figure 8.15. The stiff wire frame is connected in series with a switch, an ammeter, a variable resistor and a battery. When the switch is closed, the magnet exerts a force on the wire which can be measured from the change of the top-pan balance reading.

- **To test the variation of force with the current through the wire**, the variable resistor is adjusted to change the current. Before switching the current on, the top-pan balance reading should be noted. The top-pan balance reading is then measured for different measured values of the current. The length of the test wire in the field is kept the same. The force due to the magnetic field is worked out from the change of the top-pan balance reading. If this is in grams, the reading must be converted to kilograms then multiplied by g ($= 9.8$ m s^{-2}) to give the force. For example, if the change of the top-pan balance reading is 20.5 g, the force due to the magnetic field is 0.20 N ($= 20.5 \times 10^{-3}$ kg $\times 9.8$ m s^{-2}).

A graph of a typical set of results is shown in Figure 8.16.

- **To test the variation of force with the length of the wire**, the current is kept the same throughout by using the variable resistor as necessary. The length of the wire in the field is changed by reconnecting the wires to the frame. For each length, the reading of the top-pan balance is noted and the force due to the magnetic field is calculated. A graph of a typical set of results is shown in Figure 8.17.

- **To test the variation of the force with the angle between the wire and the magnetic field lines**, the magnet can be turned gradually. This test shows that the force is a maximum when the wire is perpendicular to the magnetic field lines.

The tests above show that the force F on the wire is proportional to:

- the current I,
- the length l of the wire.

Magnetic flux density

> **The magnetic flux density, B, is defined as the force per unit current per unit length on a current-carrying wire placed perpendicular to the field lines.**

The unit of B is the **tesla (T)**, equal to 1 N A^{-1} m^{-1}. The magnetic flux density is sometimes also referred to as the **magnetic field strength**.

For a wire of length l at right angles to a uniform magnetic field of flux density B, the force on the wire when current I passes through it is given by:

$$F = BIl$$

Worked example

A horizontal wire of length 0.050 m is in a uniform magnetic field directed vertically upwards. The wire lies along a line from North to South. When a current of 4.0 A is passed along the wire, a force of 5.6×10^{-2} N is exerted on the wire, as shown in Figure 8.18.

a Calculate the magnetic flux density of the magnetic field.

b State the direction of the force on the wire if the current in the wire was from North to South.

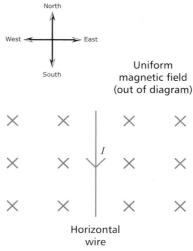

Fig 8.18

Over head view

Solution

a $B = \dfrac{F}{Il} = \dfrac{5.6 \times 10^{-2}}{4.0 \times 0.050} = 0.28$ T

b Using Fleming's left-hand rule (Fig 8.11), gives due West for the direction of the force.

8.3.2 The definition of the ampere

Two parallel current-carrying wires exert equal and opposite forces on each other, because the current in each wire creates a magnetic field that causes a force on the other wire.

As shown in Figure 8.19, the wires:

- **repel** when the currents are in opposite directions, and
- **attract** when the currents are in the same direction.

The **ampere** is defined, from this effect, as the current in two parallel conductors that causes a force of 2.0×10^{-7} N per metre length on each wire when the wires (of negligible cross-section) are 1 m apart in a vacuum. Note that knowledge of this definition is not a requirement of the specification.

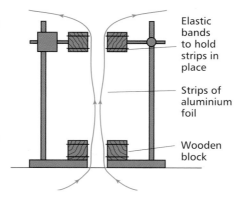

Fig 8.19 *The force between two current-carrying conductors*

QUESTIONS

1 Use Figure 8.16 to work out the magnetic flux density of the magnet that was used in the test.

2 A straight wire, of length 0.080 m, is placed vertically in a uniform magnetic field of magnetic flux density 55 mT which is horizontal and in the direction due North. A current of 4.5 A is passed up the wire. Sketch the arrangement, calculate the force on the wire and state the direction of this force.

3 Table 8.1 relates the force on a current-carrying wire, at right angles to the lines of force of a magnetic field, to the magnetic flux density and the current. Complete the table by working out the missing data in each column.

4 The Earth's magnetic field at a certain position on the Earth's surface has a horizontal component of 18 μT due North and a downwards vertical component of 55 μT. Calculate:

a the magnitude of the Earth's magnetic field at this position,

b the magnitude and direction of the force on a vertical wire of length 0.80 m carrying a current of 4.5 A downwards.

Table 8.1

	a	b	c	d
B/T	0.20 T vertically down	0.20 T vertically down	?	0.1 T horizontal due?
I/A	3.0 A horizontal due North	?	3.0 A horizontal due North	2.0 A vertically up
l/m	0.040 m	0.040 m	0.040 m	0.040 m
F/N	?	0.036 N horizontal due South	0.024 N horizontal due West	? horizontal due East

1 The diagram shows a plotting compass mid-way between a bar magnet and one end of a solenoid. The solenoid is connected in series with a battery, a variable resistor and a switch.

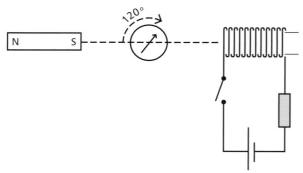

a When the switch is open, the plotting compass points directly towards the bar magnet. When the switch is closed, the needle of the plotting compass turns through 120°. Explain why the needle turns when the switch is closed.

b With the switch closed, the variable resistor is adjusted, making the needle turn back by 30°.
  (i) What must have been the effect of the adjustment of the variable resistor on the solenoid current?
  (ii) What would be the effect on the direction of the compass needle if the magnet was to be moved away from the plotting compass?

2 A plastic box containing a flat circular coil is connected to a cell in series with a switch and a variable resistor. A magnetic field is detected outside the box when the switch is closed. The pattern of this magnetic field is shown below.

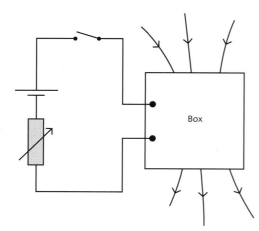

a Sketch the magnetic field pattern you would expect if the cell was reversed in the circuit.

b Sketch the magnetic field pattern you would expect to observe if the current through the coil was increased.

3 The diagram shows a long straight vertical wire connected in series with a switch, a battery and a variable resistor. When the switch is open, the needle of a plotting compass near the wire points due North.

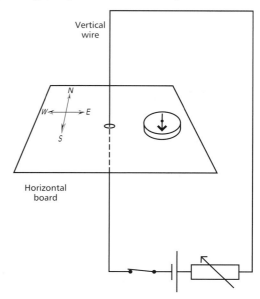

a When the switch is closed, the plotting compass needle turns and points due South. Explain why this happens.

b The current is gradually reduced using the variable resistor. Describe and explain the effect on the direction in which the needle of the plotting compass points.

c Sketch the pattern of the magnetic field close to the wire.

4 A simple electric motor consists of a rectangular coil on a spindle between opposite magnetic poles of a U-shaped magnet, as shown below.

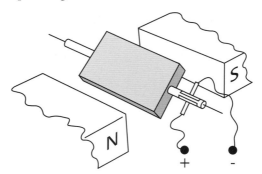

a Explain why the coil turns when the switch is closed.

b (i) Explain why the motor does not work if alternating current is used instead of direct current.

　(ii) The magnet is replaced by an electromagnet. Explain why the motor with this modification does work with alternating current.

5 A straight stiff wire is placed horizontally in a uniform horizontal magnetic field of magnetic flux density 0.25 T. The wire rests on two narrow metal supports 50 mm apart at right angles to the magnetic field, as shown below.

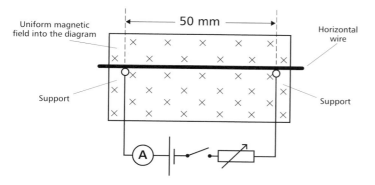

a A direct current is passed through the wire. Calculate the force on the wire when the current is 6.0 A.

b The weight of the wire is 8.0×10^{-2} N. Calculate the current in the wire that would be necessary to make the wire lift off the supports.

6 A U-shaped magnet is placed on the pan of a top-pan balance, as shown below. A straight wire is placed horizontally between the poles of the magnet, which are of length 32 mm. The wire is connected to a variable resistor, an ammeter, a switch and a battery.

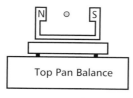

a When the switch is closed, the reading of the top-pan balance changes by 0.028 N when the ammeter reads 3.8 A. Calculate the magnetic flux density between the poles of the magnet.

b Calculate the force on the wire if the current is increased to 7.0 A.

7 The Earth's magnetic field at a certain location has a downward vertical component of 58 μT, and a horizontal component of 18 μT in a direction due North. A horizontal cable of length 50 m lies along a line from North to South. The cable carries a direct current of 26 A from South to North.

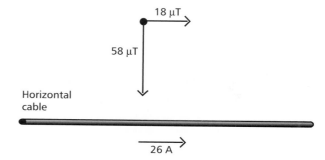

a (i) Show that the magnitude of the force on the cable due to the Earth's magnetic field is 7.5×10^{-2} N.

　(ii) State the direction of the force on the cable.

b Without further calculation, explain why the force on the cable for the same current would have been different, had the cable been aligned along a line from East to West instead of from North to South.

8 The armature coil of an electric motor has 100 turns and is of length 0.12 m as shown below. The coil spins between the poles of a U-shaped magnet, where the magnetic flux density is 0.18 T.

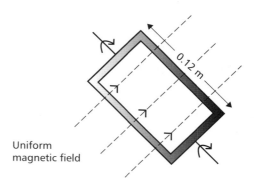

a (i) Calculate the force on each side of the coil when the current through it is 0.8 A.

　(ii) Discuss how the force on each side of the coil changes during one complete rotation of the coil.

b Explain why the force acting on each side of the coil has its maximum turning effect when the plane of the coil is parallel to the lines of force of the magnetic field.

9 Electromagnetic waves and quantum theory

9.1 The electromagnetic spectrum

9.1.1 Electromagnetic waves

Light is just a small part of the **spectrum** of electromagnetic waves. Our eyes cannot detect the other parts. The world would appear very different to us if they could. For example, all objects emit infra-red radiation; infra-red cameras enable objects to be observed in darkness.

All electromagnetic waves travel in a vacuum at the speed of light, c, which is 3.0×10^8 m s^{-1}. Therefore the wavelength, λ, of electromagnetic radiation of frequency, f, in a vacuum is given by the equation:

$$\text{Wavelength, } \lambda = \frac{c}{f}$$

The main branches of the different parts of the electromagnetic spectrum are listed in Table 9.1.

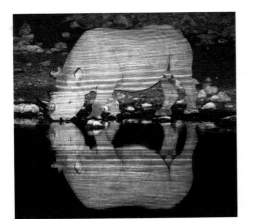

Fig 9.1 An infra-red camera in action

Table 9.1 The main branches of the electromagnetic spectrum

Type	Radio	Microwave	Infra-red	Visible	Ultra-violet	X-rays	Gamma rays
Wavelength range	>0.1 m	0.1 m to 1 mm	1 mm to 700 nm	700 nm to 400 nm	400 nm to 1 nm	<1nm	<1nm

The nature of electromagnetic waves

Electromagnetic waves were predicted by James Clerk Maxwell in 1862. Maxwell knew that a magnetic field is created round a wire when an electric current passes along the wire. He knew about Michael Faraday's discovery that a changing magnetic field in a coil of wire induces a voltage in the coil. He wondered if the two effects could be linked, and he used his mathematical skills to discover the link. In effect, Maxwell showed that the changing magnetic field created by an alternating current in a wire creates an alternating electric field, which creates an alternating magnetic field further away, which creates an alternating electric field, and so on.

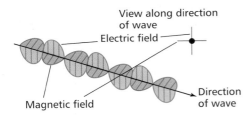

Fig 9.2 Electromagnetic waves

Maxwell showed that the result is an **electromagnetic wave** consisting of an alternating magnetic field in phase with an alternating electric field. He worked out that the speed of an electromagnetic wave should be 300 000 km s^{-1}, the same as the speed of light in a vacuum!

Maxwell realized, from his theory, that light must be an electromagnetic wave and that electromagnetic waves must exist beyond the visible spectrum. He knew that infra-red and ultra-violet radiation are outside the visible spectrum, and he correctly predicted electromagnetic waves beyond these two invisible forms of radiation.

An electromagnetic wave consists of an **alternating magnetic field** (the magnetic wave) and an **alternating electric field** (the electric wave), in which the magnetic wave and the electric wave:

- are at right angles to each other,
- vibrate in phase with each other,
- both vibrate at right angles to the direction of propagation of the electromagnetic wave.

9.1.2 Photons

Electromagnetic waves are emitted by a charged particle when it loses energy. This can happen when:

- a fast-moving electron is stopped, or
- an electron in a shell of an atom moves to a different shell of lower energy.

Electromagnetic waves are emitted as short 'bursts' of waves, each 'burst' leaving the source in a different direction. Each 'burst' is a packet of electromagnetic waves and is referred to as a **photon**. The photon theory was established by Einstein in 1905, when he used his ideas to explain **photoelectricity**. This is the emission of electrons from a metal surface when light is directed at the surface (see section 9.3). Einstein imagined photons to be like 'flying needles', and he assumed that the energy E of a photon depends on its frequency f in accordance with the equation:

Fig 9.3 Emitting photons

Photon energy, $E = hf$

where h is a constant referred to as the **Planck constant**. The value of h is 6.63×10^{-34} J s.

Worked example

$h = 6.63 \times 10^{-34}$ J s, $c = 3.00 \times 10^{8}$ m s^{-1}

Calculate the frequency and the energy of a photon of wavelength 590 nm.

Solution

To calculate the frequency, use $f = \dfrac{c}{\lambda} = \dfrac{3.00 \times 10^{8}}{590 \times 10^{-9}} = 5.08 \times 10^{14}$ Hz

To calculate the energy of a photon of this wavelength, use:

$$E = hf = 6.63 \times 10^{-34} \times 5.08 \times 10^{14} = 3.37 \times 10^{-19} \text{ J}$$

Laser power

A **laser beam** consists of photons of the same frequency. The power of a laser beam is the energy per second carried by the photons. For a beam consisting of photons of frequency f:

Power of the beam $= nhf$

where n is the number of photons per second in the beam passing a fixed point. This is because each photon has energy hf. Therefore, if n photons pass a fixed point each second, the energy passing that point each second is nhf.

Fig 9.4 A laser at work

QUESTIONS

$c = 3.00 \times 10^{8}$ m s^{-1}, $h = 6.63 \times 10^{-34}$ J s

1 **a** List the main parts of the electromagnetic spectrum in order of increasing wavelength.

   b Calculate the frequency of:
   (i) light of wavelength 590 nm,
   (ii) radio waves of wavelength 200 m.

2 With the aid of a suitable diagram, explain what is meant by an electromagnetic wave.

3 Light from a certain light source has a wavelength of 430 nm. Calculate:

   a the frequency of light of this wavelength,

   b the energy of a photon of this wavelength.

4 **a** Calculate the frequency and energy of a photon of wavelength 635 nm.

   b A laser emits light of wavelength 635 nm in a beam of power 1.5 mW. Calculate the number of photons emitted by the laser each second.

9.2.1 Beyond the visible spectrum

Radio waves and microwaves

- Radio waves were first discovered by Heinrich Hertz in 1887. Hertz showed that radio waves are produced when high voltage sparks jump across a gap. He used a wire loop with a small gap in it as a detector. When radio waves pass through the loop, tiny sparks are induced across the detector gap.

- Radio waves and microwaves are used to carry information from aerials supplied with high-frequency alternating current. The carrier waves are modulated by the information signal which is used to vary the frequency or the amplitude of the alternating current.

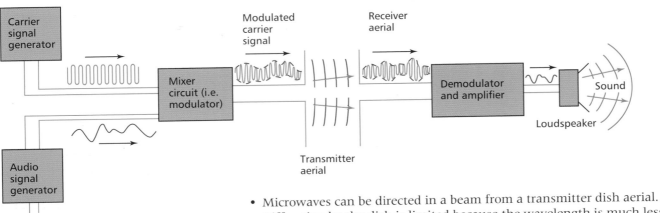
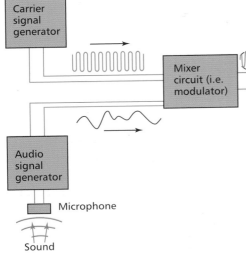

Fig 9.5 Radio communication

- Microwaves can be directed in a beam from a transmitter dish aerial. Diffraction by the dish is limited because the wavelength is much less than the dish diameter. Therefore, the beam can be directed and detected by a receiver dish along the line of sight of the beam. Microwaves travel through the atmosphere without significant absorption and, therefore, can be used for satellite communications.

- Microwaves are used for heating. Food in a microwave oven becomes hot throughout because the microwaves penetrate into the food and make the water molecules in the food vibrate.

Infra-red radiation

- Every object emits infra-red radiation. The higher the temperature of the source, the greater the amount of infra-red radiation it emits.

- Infra-red radiation can be detected by producing a visible spectrum using white light and placing a blackened thermometer bulb just beyond the red part of the visible spectrum. The thermometer shows a rise of temperature because it absorbs infra-red radiation.

- Infra-red radiation is used for imaging (Fig 9.1), for heating and drying, and for communications. Infra-red beams are used to carry data in digital form along optical fibres.

Ultra-violet radiation

- Ultra-violet radiation can be detected by placing a strip of fluorescent paper just beyond the violet part of the visible spectrum. The paper glows due to absorbing ultra-violet light. Ultra-violet marker pens contain ink that is not visible except when it absorbs ultra-violet radiation and glows.

- Ultra-violet radiation is harmful to the skin and to the eyes. Protective goggles should be worn by anyone using a sunbed. The ozone layer of the Earth's

Fig 9.6 Eye protection

atmosphere filters out most, but not all, of the ultra-violet radiation from the Sun. The thinning of the ozone layer is a cause for concern and is the reason why people should wear hats and suncream when outdoors in summer.

X-rays and gamma rays

- X-rays and gamma rays cover the same range of wavelengths. They differ in that gamma rays are produced when an atomic nucleus loses energy, whereas X-rays are emitted when electrons in a beam are suddenly stopped or when an inner shell electron in an atom moves to a shell closer to the nucleus.

- X-rays and gamma rays pass through human tissue, but are stopped by bone. When an X-ray photograph is taken, the relevant part of the patient is exposed to X-rays from an X-ray tube. A photographic film, in a light-proof wrapper, is placed in the path of the X-rays that pass through the patient. A 'negative' image of the bones is formed on a photographic film.

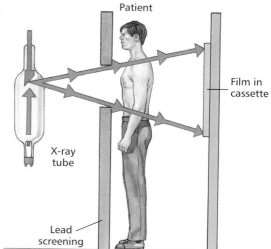

Fig 9.7 Taking a chest X-ray

9.2.2 Comparison of different types of electromagnetic wave

Table 9.2 Comparison of different types of electromagnetic wave

Type	Production	Detection	Properties
Radio	Rapid acceleration and deceleration of electrons in aerials	Receiver aerials	Reflected by metals
Microwave	Magnetron or klystron valve	Point contact diodes	Reflected by metals Absorbed partly by non-metals
Infra-red	Vibrations of atoms and molecules	Infra-red sensors Infra-red cameras	Polished silvery surfaces are the best reflectors Matt black surfaces are the best emitters
Visible	Atomic electrons when they move from one energy level to a lower level	The eye Photocells Photographic film	In white light, coloured surfaces only reflect their own colour of light All other colours are absorbed
Ultra-violet	Inner shell electrons moving from one energy level in an atom to a lower level	Photocells Photographic film	Absorbed by glass
X-rays	X-ray tubes or inner shell electron transitions	Photographic film Geiger tube Ionisation chamber	Penetrate metal Ionise gases
Gamma rays	Radioactive decay of the nucleus	Same as X-rays	Same as X-rays

Electromagnetic waves and living cells

Radio waves and microwaves penetrate tissues and are absorbed by the water content of living cells, causing internal heating which may damage or even destroy the cell.

Infra-red radiation is absorbed by skin cells, causing internal heating which may damage or destroy the cells.

Ultra-violet radiation damages the cells of the retina of the eye. Special protective goggles must be worn by users of sunbeds. UV radiation also damages cells below the skin, because it can penetrate more than infra-red radiation. Users of sunbeds must **not** exceed recommended exposure times or severe burns will result.

X-rays and gamma radiation create ions in substances which they pass through. Living cells are damaged by ionising radiation. High doses kill cells and low doses can cause cell mutation and cancerous growth. There is no evidence for a lower limit below which living cells would not be damaged. The maximum permissible radiation dose is a legal limit decided on the basis of acceptable risk.

QUESTIONS

1 a Why must sunbed users wear protective goggles?
 b Why is the thinning of the Earth's ozone layer a matter of great concern?

2 a Explain why an infra-red TV camera can be used to observe the movement of animals at night.
 b Explain why a hot object wrapped in silver foil cools more slowly than a similar unwrapped object.

3 a State two types of electromagnetic wave used for:
  (i) communications,
  (ii) heating.
 b Give two reasons why microwaves, rather than radio waves, are used for satellite communications.

4 a State the difference between X-rays and gamma rays.
 b State one property of X-rays and gamma rays not shared with other forms of electromagnetic radiation.

9.3.1 The discovery of photoelectricity

A metal contains conduction electrons which move about freely inside the metal. These electrons collide with each other and with the positive ions of the metal. Heinrich Hertz discovered how to produce and detect radio waves (see section 9.2). He found that the sparks produced in his spark gap detector, when radio waves were being transmitted, were stronger when ultra-violet radiation was directed at the spark gap. Further investigations of the effect of electromagnetic radiation on metals showed that electrons are emitted from the surface of a metal when electromagnetic radiation above a certain frequency is directed at the metal. This effect is known as the **photoelectric effect**.

Demonstration of the photoelectric effect

Ultra-violet radiation from a UV lamp is directed at the surface of a zinc plate placed on the cap of a gold leaf electroscope, as shown in Figure 9.8. This device is a very sensitive detector of charge. When it is charged, the thin gold leaf of the electroscope rises; it is repelled from the metal 'stem', because they are both charged with a like charge.

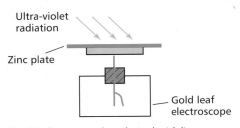

Ultra-violet radiation

Zinc plate

Gold leaf electroscope

Fig 9.8 *Demonstrating photoelectricity*

- If the electroscope is **charged negatively**, the leaf rises and stays in position. However, if ultra-violet light is directed at the zinc plate, the leaf gradually falls. The leaf falls because conduction electrons at the zinc surface leave the zinc surface when ultra-violet light is directed at it. The emitted electrons are referred to as **photoelectrons**.

- If the electroscope is **charged positively**, the leaf rises and stays in position, regardless of whether or not ultra-violet light is directed at the zinc plate. This is because the conduction electrons in the zinc plate are held on the zinc plate, as the plate is charged positive and electrons carry negative charge.

Puzzling problems

The following observations were made about photoelectricity after Hertz's discovery. These observations were a major problem, because they could not be explained using the wave theory of electromagnetic radiation.

- Photoelectric emission of electrons from a metal surface does not take place if the frequency of the incident electromagnetic radiation is below a certain value, known as the **threshold frequency**. This minimum frequency depends on the type of metal.
  Note The **wavelength** of the incident light must be less than a **maximum** value equal to the speed of light/the threshold frequency.

- The number of electrons emitted per second is proportional to the **intensity** of the incident radiation, provided the frequency is greater than the threshold frequency. If the frequency of the incident radiation is less than the threshold frequency, no photoelectric emission from that metal surface can take place, no matter how intense the incident radiation is.

- Photoelectric emission occurs without delay, as soon as the incident radiation is directed at the surface, provided the frequency of the radiation exceeds the threshold frequency. No matter how weak the intensity of the incident radiation is, electrons are emitted as soon as the source of radiation is switched on.

The wave theory of light cannot explain either the existence of a threshold frequency, or why photoelectric emission occurs without delay. According to wave theory, each conduction electron at the surface of a metal should gain some energy from the incoming waves, regardless of how many waves arrive each second. Wave theory therefore predicted that:

- Emission should take place with waves of any frequency.
- Emission would take longer using low intensity waves than using high intensity waves.

The discovery of radio waves and X-rays confirmed the prediction by Maxwell (section 9.1) that electromagnetic radiation exists beyond the known spectrum from ultra-violet to infra-red radiation. Using Maxwell's theory of electromagnetic waves, physicists were very successful in predicting the properties of electromagnetic waves until the discovery of the photoelectric effect.

9.3.2 Einstein's explanation of photoelectricity

The photon theory of light was put forward by Einstein in 1905 to explain photoelectricity. As explained in section 9.1, Einstein assumed that light is composed of wave packets, or **photons**, each of energy equal to hf, where f is the frequency of the light and h is the Planck constant. The accepted value for h is 6.63×10^{-34} J s:

Energy of a photon = hf

For electromagnetic waves of wavelength λ, the energy of each photon $E = hf = \dfrac{hc}{\lambda}$

where c is the speed of the electromagnetic waves.

To explain photoelectricity, Einstein said that:

- When light is incident on a metal surface, an electron at the surface absorbs a **single** photon from the incident light and therefore gains energy equal to hf, where hf is the energy of a light photon.
- An electron can leave the metal surface, if the energy gained from a single photon exceeds **the work function, ϕ,** of the metal. This is the minimum energy needed by an electron to escape from the metal surface.

Hence the maximum kinetic energy of an emitted electron:

$$E_{Kmax} = hf - \phi$$

Emission can take place from a surface at zero potential, provided $E_{Kmax} > 0$, i.e. $hf > \phi$. Thus the threshold frequency of the metal:

$$f_{min} = \frac{\phi}{h}$$

Fig 9.9 Albert Einstein 1879–1955

In 1905 Einstein published three papers, each of which changed existing ideas. His paper on Brownian motion showed that atoms must exist, and he established the photon theory of light in the second paper. Perhaps he is best remembered for his theories of relativity which he stated in the third paper.

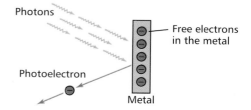

Fig 9.10 Explaining photoelectricity

QUESTIONS

$h = 6.63 \times 10^{-34}$ J s, $c = 3.00 \times 10^8$ m s^{-1}

1 a What is meant by 'photoelectric emission' from a metal surface?

 b Explain why photoelectric emission from a metal surface only takes place if the frequency of the incident radiation is greater than a certain value.

2 a Calculate the frequency and energy of a photon of wavelength:
 (i) 450 nm (ii) 1500 nm

 b A metal surface at zero potential emits electrons from its surface if light of wavelength 450 nm is directed at it, but not if light of wavelength 650 nm is used. Explain why photoelectric emission happens with light of wavelength 450 nm but not with light of wavelength 650 nm.

3 The work function of a certain metal plate is 1.1×10^{-19} J. Calculate:

 a the threshold frequency of incident radiation,

 b the maximum kinetic energy of photoelectrons emitted from this plate when light of wavelength 520 nm is directed at the metal surface.

4 Light of wavelength 635 nm is directed at a metal plate at zero potential. Electrons are emitted from the plate with a maximum kinetic energy of 1.5×10^{-19} J. Calculate:

 a the energy of a photon of this wavelength,

 b the work function of the metal,

 c the threshold frequency of electromagnetic radiation incident on this metal.

9.4.1 More about conduction electrons

The **average kinetic energy** of a conduction electron in a metal depends on the **temperature** of the metal. As the conduction electrons move about at random in the metal, they can be likened to the molecules of a gas. The average kinetic energy of a gas molecule is proportional to the absolute temperature of the gas. It can be shown that the average kinetic energy of an conduction electron in a metal at 300 K is therefore about 6×10^{-21} J.

- The **work function** of a metal is the **minimum** energy needed by a conduction electron to escape from the metal surface when the metal is at zero potential. The work function of a metal is of the order of 10^{-19} J, which is about 20 times greater than the average kinetic energy of a conduction electron in a metal at 300 K. In other words, a conduction electron in a metal at about 20 °C does not have sufficient kinetic energy to leave the metal.

- When a conduction electron **absorbs a photon**, its kinetic energy increases by an amount equal to the energy of the photon. Provided the energy of the photon exceeds the work function of the metal, the conduction electron can leave the metal. If the electron does not leave the metal, it collides repeatedly with other electrons and positive ions, and it quickly loses its extra kinetic energy.

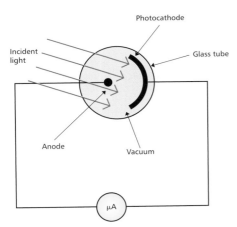

Fig 9.11 *Using a vacuum photocell*

The electron volt

The electron volt (eV) is a unit of energy equal to the work done when an electron is moved through a p.d. of 1 V.

For a charge q moved through a p.d., V:

$$\text{Work done} = qV$$

Therefore, the work done when an electron moves through a potential difference of 1 V is equal to **1.6×10^{-19} J** ($= 1.6 \times 10^{-19}$ C $\times$ 1 V). This amount of energy is defined as **1 electron volt.**

Examples

The work done on:

- an electron when it moves through a potential difference of 1000 V = 1000 eV,
- an ion of charge $+2e$ when it moves through a potential difference of 10 V = 20 eV.

9.4.2 Photoelectricity investigations

The vacuum photocell

A **vacuum photocell** is a glass tube that contains a metal plate, referred to as the **photocathode**, and a smaller metal electrode referred to as the **anode**. Figure 9.11 shows a vacuum photocell in a circuit. When light of frequency greater than the threshold frequency for the metal is directed at the photocathode, electrons emitted from the cathode transfer to the anode. The micro ammeter in the circuit can be used to measure the photoelectric current, which is proportional to the number of electrons per second that transfer from the cathode to the anode.

- For a photoelectric current I, the **number of photoelectrons per second** that transfer from the cathode to the anode is I/e, where e is the charge of the electron.

- The photoelectric current is proportional to the **intensity** of the light incident on the cathode. This is because the intensity of the incident light is a measure of the **energy per second** carried by the incident light, which is proportional to the number of photons per second incident on the cathode. Because each photoelectron must have absorbed one photon to escape from the metal surface, the number of photoelectrons emitted per second (i.e. the photoelectric current) is therefore proportional to the intensity of the incident light.

- The intensity of the incident light does **not** affect the maximum kinetic energy of a photoelectron. No matter how intense the incident light is, the energy gained by a photoelectron is due to the absorption of one photon only. Therefore, the maximum kinetic energy of a photoelectron is still given by $E_{\text{Kmax}} = hf - \phi$, as explained in section 9.3.

Measurement of the work function of a metal

Photoelectric emission can be stopped by making the photocathode sufficiently positive. Figure 9.12 shows how a potential divider can be used to make the photocathode increasingly positive, relative to the anode. As the potential difference is increased from zero, the micro-ammeter reading decreases to zero. This happens because each photoelectron leaving the metal surface needs to do extra work, as the plate is at a positive potential. The kinetic energy of a photoelectron is therefore reduced, because each electron has to do work to overcome the attraction of the plate. When the micro ammeter reading is zero, photoelectric emission stops; this is because the kinetic energy of a photoelectron is reduced to zero before the electron can escape from the attraction of the plate.

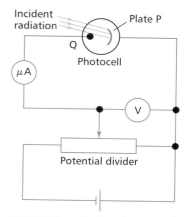

- As explained earlier, the work done by a charged particle when it moves through a potential difference V is qV, where q is the charge of the particle. Therefore, to escape from a metal plate at positive potential V, the **extra work** needed to be done by a photoelectron is eV, where e is the charge of the electron.

- At zero potential, the **maximum kinetic energy** of an emitted photoelectron is $hf - \phi$, where f is the frequency of the incident radiation and ϕ is the work function of the metal.

Fig 9.12 *Investigating photoelectricity*

Therefore, photoelectric emission is stopped when the potential V is such that:

$$eV_S = hf - \phi$$

where V_S is the potential needed to stop emission.

In other words, the maximum kinetic energy of a photoelectron from a surface at zero potential:

$$E_{Kmax} = eV_S$$

By measuring V_S for different frequencies f, E_{Kmax} can be calculated for each frequency; then a graph of E_{Kmax} against frequency f can be plotted. Because $E_{Kmax} = hf - \phi$, the graph is a straight line (see p.157) with:

- a gradient h,
- a y-intercept equal to $-\phi$, and
- an x-intercept equal to the threshold frequency, $f_{min} = \dfrac{\phi}{h}$

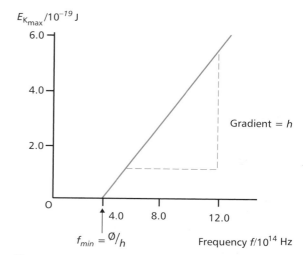

Fig 9.13 *A graph of E_{Kmax} against frequency*

QUESTIONS

$h = 6.63 \times 10^{-34}$ J s, $c = 3.00 \times 10^8$ m s^{-1}, $e = 1.6 \times 10^{-19}$ C

1 A vacuum photocell is connected to a micro ammeter. Explain the following observations:

   a When the cathode was illuminated with blue light of low intensity, the micro ammeter showed a non-zero reading.

   b When the cathode was illuminated with an intense red light, the micro ammeter reading was zero.

2 A vacuum photocell is connected to a micro ammeter. When light is directed at the photocell, the micro ammeter reads 0.25 μA.

   a Calculate the number of photoelectrons emitted per second by the photocathode of the photocell.

   b Explain why the micro ammeter reading is doubled if the intensity of the incident light is doubled.

3 A narrow beam of light of wavelength 590 nm and of power 0.5 mW is directed at the photocathode of a vacuum photocell, which is connected to a micro-ammeter that reads 0.4 μA. Calculate:

   a the energy of a single light photon of this wavelength,

   b the number of photons per second incident on the photocathode,

   c the number of electrons emitted per second from the photocathode.

4 a Use Figure 9.13 to estimate:
   (i) the threshold frequency,
   (ii) the work function of the photocathode that gave the results used to plot the graph.

   b A metal surface has a work function of 1.9×10^{-19} J. Light of wavelength 435 nm is directed at the metal surface. Calculate the maximum kinetic energy of the photoelectrons emitted from this metal surface.

Wave particle duality

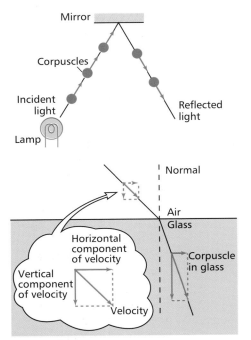

Fig 9.14 Newton's corpuscular theory of light

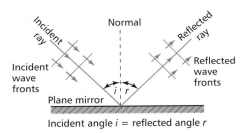

Fig 9.15 Huygens' wave theory of light

Incident angle *i* = reflected angle *r*

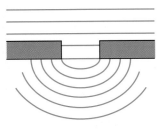

Fig 9.16 Diffraction

9.5.1 The nature of light

Newton v. Huygens

Light was considered by Newton to be composed of tiny particles, which he called **corpuscles**. He explained the law of reflection of light by imagining the corpuscles bounced off the mirror like a squash ball bounces off a wall. He explained the refraction of light in glass or water by assuming the corpuscles travel faster in glass or water than they do in air, as shown in Figure 9.14.

Huygens put forward the **wave theory of light** as an alternative to Newton's theory of light. As explained in section 11.1, reflection and refraction may be explained using the wave theory of light. To explain **refraction** using wave theory, it is necessary to assume that light travels slower in glass or water than it does in air. Huyghens' wave theory of light was rejected in favour of Newton's corpuscular theory, because:

- There was no evidence, at the time, that light possessed wave properties such as diffraction.

- It was not known, at the time, if light travelled slower or faster in glass or water than it does in air.

Most scientists considered Newton's theory to be correct, because Newton's scientific reputation was much stronger. The discovery of interference of light by Young, over a century later, made some scientists think of light in terms of waves. However, the wave theory was not finally accepted until a few decades later, when the speed of light was shown to be less in water than in air.

The dual nature of light

As explained in section 9.1, Maxwell showed that light is part of the electromagnetic spectrum of waves. The theory of electromagnetic waves predicted the existence of electromagnetic waves beyond the visible spectrum. The subsequent discovery of X-rays and radio waves confirmed the predictions and seemed to show that the nature of light had been settled. Many scientists in the late nineteenth century reckoned that all aspects of physics could be explained using Newton's laws of motion and Maxwell's theory of electromagnetic waves. They thought that the few minor problem areas, such as photoelectricity, would be explained sooner or later using Newton's laws of motion and Maxwell's theory of electromagnetic waves. However, as explained in section 9.3, photoelectricity could not be explained until Einstein put forward the radical theory that light consists of photons which are packets of electromagnetic waves.

- According to Maxwell, electromagnetic waves spread out and become less intense as they do so.

- According to Einstein, electromagnetic waves are emitted in wavepackets. Each wavepacket travels without spreading out and carries a certain amount of energy. A photon is the **quantum**, or smallest amount of light possible. Light has a **dual nature**, in that it can behave as a wave or as a particle according to circumstances.

The wave-like nature is observed when diffraction or interference of light takes place:

- **Diffraction** takes place when light passes through a narrow slit. The light emerging from the slit spreads out in the same way as water waves spread out after passing through a gap. The narrower the gap, or the longer the wavelength, the greater the amount of diffraction.

• **Interference** takes place where diffracted light from two closely-spaced narrow slits overlaps. An interference pattern of bright and dark fringes is observed. This effect is explained in more detail in section 11.4.

The particle-like nature is observed, for example, in the photoelectric effect. Emission of electrons from a metal surface occurs if the frequency of the incident light is greater than the threshold frequency for that metal. If an electron near the surface absorbs a photon of frequency f, the kinetic energy of the electron is increased by hf from a negligible value. As outlined in section 9.3, the electron can only escape if the energy it gains from a photon exceeds the work function of the metal. This condition therefore explains why the frequency must exceed the threshold frequency.

9.5.2 Matter waves

If light has a dual wave particle nature, perhaps **particles of matter** also have a dual nature. This interesting question was first considered by de Broglie in 1923. By extending the ideas of duality from photons to matter particles, de Broglie put forward the hypothesis that matter particles ought to have a dual wave particle nature. He also suggested that the wave-like behaviour of a matter particle is characterised by a wavelength, its de Broglie wavelength, λ_{db}, which he related to the momentum, p, of the particle by means of the equation:

$$\lambda_{db} = \frac{h}{p}$$

where h is the Planck constant.

The momentum of a moving particle is defined as its mass $\times$ its velocity. So a particle of mass m moving at velocity v has a de Broglie wavelength given by:

$$\lambda_{db} = \frac{h}{mv}$$

Evidence for de Broglie's theory of matter waves was discovered three years later, when it was demonstrated that a beam of electrons can be diffracted. Figure 9.17 shows how this is done. After this discovery, further experimental evidence using other types of particle confirmed the correctness of de Broglie's theory.

• The beam of electrons is produced in a **vacuum tube**, by attracting electrons from a heated filament wire towards a positively charged anode. A hole in the anode allows some electrons through into a beam. The electrons emerge from the beam at a speed that depends on the p.d. between the filament and the anode. The greater the p.d., the faster the electrons in the beam.

• The beam is directed at a **thin metal foil**. A metal is composed of many tiny crystalline regions. Each region, or 'grain', consists of positive ions arranged in fixed positions in rows in a regular pattern. The rows of atoms cause the electrons in the beam to be diffracted, just as a beam of light is diffracted by the lines of a diffraction grating.

• The electrons in the beam pass though the metal foil and are **diffracted** in certain directions only, as shown in Figure 9.17. They form a pattern of rings on a fluorescent screen at the end of the tube. Each ring is due to electrons diffracted by the same amount from grains at the same angle to the incident beam.

• If the p.d. between the filament and the anode is increased, the **speed of the electrons** is increased. As a result, the diffraction rings become smaller. This is because the increase of speed makes the de Broglie wavelength smaller, so less diffraction occurs and the rings become smaller.

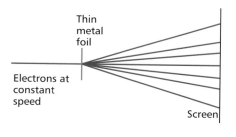 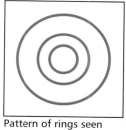

Pattern of rings seen on the screen

Fig 9.17 *Diffraction of electrons*

QUESTIONS

$h = 6.6 \times 10^{-34}$ J s,
mass of an electron $= 9.1 \times 10^{-31}$ kg,
mass of a proton $= 1.7 \times 10^{-27}$ kg

1 With the aid of examples, explain what is meant by the dual wave particle nature of:

  a light,

  b matter particles.

2 State whether each of the following experiments demonstrates the wave nature or the particle nature of matter or of light:

  a photoelectricity,

  b electron diffraction.

3 Calculate the de Broglie wavelength of:

  a an electron moving at a speed of 2.0×10^7 m s^{-1},

  b a proton moving at the same speed.

4 Calculate the momentum and speed of:

  a an electron that has a de Broglie wavelength of 500 nm,

  b a proton that has the same de Broglie wavelength.

1 a (i) List the following parts of the electromagnetic spectrum in order of increasing wavelength:

         infra-red radiation   microwaves
         ultra-violet radiation   visible light
              X-rays

     (ii) State one part of the electromagnetic spectrum not listed and indicate where on your list it should be placed.

b State two properties common to all types of electromagnetic radiation.

2 In a fluorescent tube, ultra-violet radiation produced inside the tube is absorbed by a coating of fluorescent material on the inside surface of the tube. Light is emitted by the coating material as a result:

a (i) Calculate the energy of a photon of wavelength 100 nm.

     (ii) State which part of the electromagnetic spectrum such a photon is in.

b (i) Calculate the energy of a photon of light of wavelength 500 nm.

     (ii) How many photons of wavelength 500 nm would have the same energy as one photon of wavelength 100 nm?

3 Electrons can escape from a certain metal surface when it is illuminated by blue light, but not when it is illuminated by red light:

a Explain why blue light causes emission of electrons from this metal, whereas red light does not.

b What difference would be made to the emission of electrons from this metal surface if:

    (i) the intensity of the blue light were increased,

    (ii) the surface were at a positive potential?

4 Light of wavelength 600 nm was used to illuminate a metal surface. The maximum kinetic energy of an electron emitted as a result was 1.3×10^{-19} J. Calculate:

a the energy of a photon of this wavelength,

b the work function of the metal surface.

5 a Explain the meaning of the term 'work function' of a metal surface.

b (i) Calculate the wavelength of a photon of energy 2.0×10^{-19} J.

     (ii) Explain why photons of wavelength longer than the value calculated in part (i) could not cause photoelectric emission from a metal surface which has a work function of 2.0×10^{-19} J.

6 A vacuum photocell connected to a micro-ammeter is shown below.

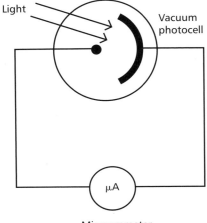

a Explain why:

    (i) the micro ammeter registers a current when light of a certain wavelength is directed at the photocell,

    (ii) the current increases when the light intensity is increased.

b When light of a longer wavelength is directed at the photocell, no current is registered:

    (i) Explain why there is no current in this situation.

    (ii) Explain why making the light more intense would not cause a current with light of this longer wavelength.

7 a State whether each of the following experiments or observations using light demonstrates the wave-like or the particle-like nature of light:

    (i) When light passes through a narrow gap, it diffracts.

(ii) When light is directed at a metal surface, electrons are emitted from the surface only if the light wavelength is less than a certain value.

b State whether each of the following experiments or observations using a beam of electrons demonstrates the wave-like or the particle-like nature of matter particles:

(i) When electrons pass through a magnetic field, they are deflected.

(ii) When electrons are directed at a thin metal sample, they are deflected in certain directions only.

8 a An electron in a beam has a speed of 1.5×10^7 m s^{-1}. Calculate:

(i) the kinetic energy,

(ii) the momentum of this electron.

b Calculate the wavelength of a photon with the same energy as the electron in part **a**.

9 a An electron in a beam has a speed of 5.0×10^6 m s^{-1}. Calculate:

(i) the kinetic energy of the electron,

(ii) the de Broglie wavelength of the electron.

b The electron in part **a** is suddenly stopped by an impact with a solid object. A photon of electromagnetic radiation is produced as a result. Assuming all the kinetic energy of the electron is used to produce a single photon, calculate the wavelength of such a photon.

10 a Outline one piece of experimental evidence that shows electrons have a wave-like nature.

b (i) Calculate the de Broglie wavelength of an electron that has a momentum of 1.7×10^{-20} kg m s^{-1}

(ii) Calculate the energy of a photon that has the same wavelength as the electron in part (i).

10 Waves

10.1 Waves and vibrations

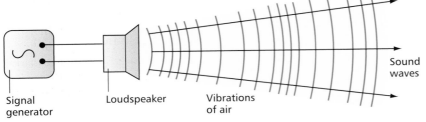

Fig 10.1 Creating sound waves in air

10.1.1 Types of wave

Waves that pass through a substance are **vibrations** of the particles of the substance. For example, sound waves in air are created by a surface that vibrates, sending compression waves through the surrounding air. Sound waves, seismic waves and waves on strings are examples of waves that pass through a substance. These types of wave are often referred to as **mechanical waves**. When waves progress through a substance, the particles of the substance vibrate in a certain way which makes other particles vibrate in the same way, and so on.

• **Electromagnetic waves** are vibrating electric and magnetic fields that progress through space, without the need for a substance. The vibrating electric field generates a vibrating magnetic field, which generates a vibrating electric field further away, and so on. Electromagnetic waves include radio waves, microwaves, infra-red radiation, light, ultra-violet radiation, X-rays and gamma radiation. The full spectrum of electromagnetic waves is listed in section 9.1.

10.1.2 Longitudinal and transverse waves

Longitudinal waves

Longitudinal waves are waves in which the direction of vibration is **parallel** to (i.e. along) the direction in which the wave travels. Sound waves, primary seismic waves and compression waves on a slinky are all longitudinal waves. Figure 10.2 shows how to send longitudinal waves along a 'slinky'. When one end of the slinky is moved to and fro repeatedly, each 'forward' movement causes a compression wave to pass along the slinky as the coils push into each other. Each 'reverse' movement causes the coils to move apart, so an 'expansion' wave passes along the slinky.

• **Sound waves** in air are created when a surface in contact with air vibrates. When the vibrating surface pushes on the air, the air molecules near the surface are pushed away from the surface, pushing on adjacent molecules which push on adjacent molecules. This creates a wave of 'high density' air (i.e. a compression) that passes through the air. When the vibrating surface 'retreats', the air molecules near the surface move back into the space vacated by the surface, allowing adjacent air molecules to fall back and so on. This creates a wave of 'low density' air (referred to as a 'rarefaction') that passes

Fig 10.2 Longitudinal waves on a slinky

through the air behind the compression wave. The vibrating surface therefore creates a series of **compressions** and **rarefactions** that pass through the air. The air molecules are therefore repeatedly pushed to and fro along the direction in which the sound travels.

Transverse waves

Transverse waves are waves in which the direction of vibration is **perpendicular** to the direction in which the wave travels. Electromagnetic waves, secondary seismic waves and waves on a string or a wire are all transverse waves. Figure 10.3 shows transverse waves travelling along a rope. When one end of the rope is moved from side to side repeatedly, these sideways movements travel along the rope: each unaffected part of the rope is pulled sideways when the part next to it moves sideways.

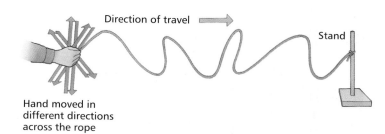

Fig 10.4 Unpolarised waves on a rope

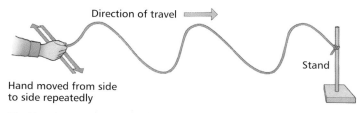

Fig 10.3 Making rope waves

- Transverse waves are **polarised** if the vibrations are in one plane only. If the vibrations change from one plane to another, the waves are **unpolarised**. Longitudinal waves cannot be polarised.

- Figure 10.4 shows **unpolarised** waves being created on a rope. Contrast this with the waves in Figure 10.3, which are polarised because they vibrate in one plane only.

- Light from a filament lamp or a candle is unpolarised. If unpolarised light is passed through a **polaroid filter**, the transmitted light is polarised: the filter only allows through light which vibrates in a certain direction, according to the alignment of its molecules. This is like using a 'letter box' in a board to polarise unpolarised waves on a rope, as shown in Figure 10.5.

- If unpolarised light is passed through two polaroid filters, the transmitted light intensity changes if one polaroid is turned relative to the other one. The filters are said to be 'crossed' when the transmitted intensity is a minimum. At this position, the polarised light from the first filter cannot pass through the second filter, as the alignment of molecules in the second filter is at 90° to the alignment in the first filter. This is like passing rope waves through two 'letter boxes' at right angles to each other, as shown in Figure 10.5.

- Light is part of the spectrum of electromagnetic waves (see section 9.1). The **plane of polarisation** of an electromagnetic wave is defined as the plane in which the electric field oscillates.

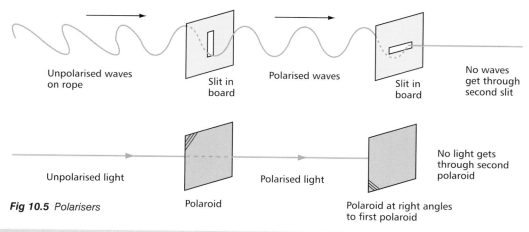

Fig 10.5 Polarisers

QUESTIONS

1 Classify the following types of wave as either longitudinal or transverse:

    radio waves   sound waves
    secondary seismic waves   microwaves

2 Sketch a snapshot of a longitudinal wave travelling on a slinky coil, indicating the direction in which the waves are travelling and areas of compression and expansion.

3 Sketch a snapshot of a transverse wave travelling along a rope, indicating the direction in which the waves are travelling and the direction of motion of the particles at the peaks and troughs.

4 a What is meant by a 'polarised' wave?

   b A light source is observed through two pieces of polaroid. Describe and explain what you would expect to observe when one of the polaroids is rotated through 360°.

Measuring waves

Fig 10.6 *Measuring electrical waves*

When an intercontinental telephone call is made, sound waves are converted to electrical waves. These waves are carried by electromagnetic waves from ground transmitters to satellites in space and back to receivers on Earth, where they are converted back to electrical waves, then back to sound waves. The electronic circuits ensure that these sound waves are very similar to the original sound waves. The engineers who design and maintain communications systems need to measure the different types of wave at different stages to make sure the waves are not distorted.

10.2.1 Key terms

The following terms, illustrated in Figure 10.7, are used to describe waves:

- The **displacement** of a vibrating particle is its distance and direction from its equilibrium position.
- The **amplitude** of a wave is the maximum displacement of a vibrating particle. For a transverse wave, this is the height of a wave crest or the depth of a wave trough from the middle.
- The **wavelength** of a wave is the least distance between two adjacent vibrating particles with the same displacement and velocity at the same time (e.g. distance between adjacent crests).
- **One complete cycle of a wave** is from a maximum displacement to the next maximum displacement (e.g. from one wave peak to the next).
- The **period** of a wave is the time for one complete wave to pass a fixed point.
- The **frequency** of a wave is the number of cycles of vibration of a particle per second. The unit of frequency is the **hertz (Hz)**. For waves of frequency f:

$$\textbf{Period of the wave} = \frac{1}{f}$$

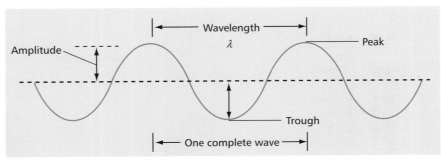

Fig 10.7 *Parts of a wave*

Using an oscilloscope to measure sound waves

The oscilloscope shown in Figure 10.8 is connected to a microphone, which detects sound waves emitted by a loudspeaker connected to a signal generator. The trace on the oscilloscope screen is the **waveform** of the sound waves produced by the loudspeaker. Figure 10.8 shows how to measure the amplitude and the frequency of this waveform using the oscilloscope controls. Note that the waveform **amplitude** is the distance **from the middle to the top**.

Wave speed

The higher the frequency of a wave, the shorter its wavelength. For example, if waves are sent along a rope, the higher the frequency at which they are produced, the closer together their wave peaks. The same effect can be seen in a ripple tank, when straight waves are produced at a constant frequency. If the

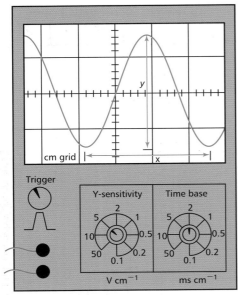

(i) Trace height y = 32 mm
∴ Trace amplitude = 16 mm = 1.6 cm
Given Y-sensitivity = 5 V cm^{-1}
Voltage amplitude = 5 × 1.6 = 8.0 volts

(ii) x-distance from peak to peak = 32 mm = 3.2 cm
Given time base 2 ms cm^{-1}
Time period = 2 × 3.2 = 6.4 ms

Frequency $f = \dfrac{1}{\text{Time period}} = \dfrac{1}{6.4 \times 10^{-3}\text{ s}}$

$f = 1.55 \times 10^3$ Hz

Fig 10.8 *Measuring a waveform*

frequency is raised to a higher value, the waves are closer together. Consider Figure 10.9, which represents the crests of straight waves in a ripple tank travelling at constant speed.

- Each wave crest travels a distance equal to one wavelength (λ) in the time for one cycle.
- The time taken for one cycle $= \dfrac{1}{f}$, where f is the frequency of the waves.

Therefore, the speed of the waves:

$$v = \frac{\text{distance travelled in one cycle}}{\text{time taken for one cycle}} = \frac{\lambda}{1/f} = f\lambda$$

For waves of frequency f and wavelength λ:

$$\textbf{Wave speed, } v = f\lambda$$

Note The symbol c is used for the speed of electromagnetic waves.

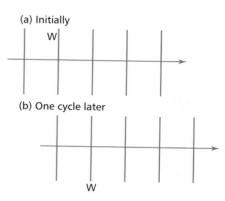

Fig 10.9 Wave speed

10.2.2 Phase difference

The **phase difference** between two vibrating particles is the fraction of a cycle between the vibrations of the two particles, measured either in **degrees or radians** where:

$$1 \text{ cycle} = 360° = 2\pi \text{ radians}$$

For two points at distance d apart along a wave of wavelength λ,

$$\textbf{Phase difference, in radians} = \frac{2\pi d}{\lambda}$$

Figure 10.10 shows three successive snapshots of the particles of a transverse wave that progresses from left to right across the diagram. Particles O, P, Q, R and S are spaced approximately $\frac{1}{4}$ of a wavelength apart. Table 10.1 shows the phase difference between O and each of the other particles.

Table 10.1 Phase differences

		P	Q	R	S
Distance from O		$\frac{1}{4}\lambda$	$\frac{1}{2}\lambda$	$\frac{3}{4}\lambda$	λ
Phase difference relative to O/radians		$\frac{1}{2}\pi$	π	$\frac{3}{2}\pi$	2π

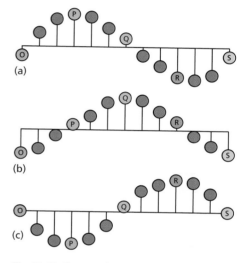

Fig 10.10 Progressive waves

QUESTIONS

1 Sound waves, in air, travel at a speed of 340 m s⁻¹ at 20 °C. Calculate the wavelength of sound waves, in air, which have a frequency of:

 a 3400 Hz

 b 18 000 Hz

2 Electromagnetic waves, in air, travel at a speed of 3.0×10^8 m s⁻¹. Calculate the frequency of electromagnetic waves of wavelength:

 a 0.030 m

 b 600 nm

3 Figure 10.11 shows a waveform on an oscilloscope screen, when the y-sensitivity of the oscilloscope was 0.50 V cm⁻¹ and the time base was set at 0.5 ms cm⁻¹. Determine the amplitude and the frequency of this waveform.

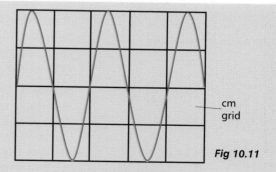

Fig 10.11

4 For the waves in Figure 10.10:

 a determine:

 (i) the amplitude and the wavelength,

 (ii) the phase difference between P and R,

 (iii) the phase difference between P and S.

 b what would be the displacement and direction of motion of Q three-quarters of a cycle after the last snapshot?

10.3 Wave properties 1

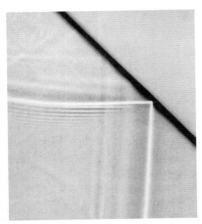

Stroboscope

Lamp

Each wave crest acts like a convex lens and concentrates the light onto the screen. So the pattern on the screen shows the wave crests.

Light

Screen

Wavecrest

Water in ripple tank

Pattern of wave fronts 'cast' on the screen

Fig 10.12 *The ripple tank*

Wave properties, such as reflection, refraction and diffraction, occur with many different types of wave. A **ripple tank** may be used to study these wave properties. The tank is a shallow transparent tray of water with sloping sides. The slopes prevent waves reflecting off the sides of tank. If they did reflect, it would be difficult to see the waves to be observed.

- The waves observed in a ripple tank are referred to as '**wave fronts**', which are lines of constant phase (e.g. crests).
- The direction in which a wave travels is at right angles to the wave front.

10.3.1 Reflection

- **Straight waves** directed at a certain angle to a hard flat surface (the '**reflector**') reflect off at the same angle, as shown in Figure 10.13. The angle between the reflected wave front and the surface is the **same** as the angle between the incident wave front and the surface. Therefore the direction of the reflected wave is at the same angle to the reflector as the direction of the incident wave. This same effect is observed when a light ray is directed at a plane mirror. The angle between the incident ray and the mirror is the **same** as the angle between the reflected ray and the mirror.
- **Circular waves** from a point source spread out equally in all directions. Figure 10.14 shows what happens when circular waves hit a flat reflector. The reflected waves appear to come from a point image the same distance behind the reflector as the point source is in front of it. This is why the image of an object observed in a plane mirror is the same distance behind the mirror as the object is in front of the mirror. Reflected light from each point of the object appears to come from a corresponding **image point** behind the mirror at the same distance.

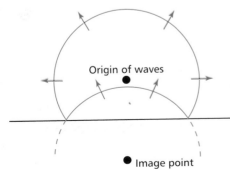

Fig 10.13 *Reflection of plane waves*

10.3.2 Refraction

When waves pass across a boundary at which the wave speed changes, the wavelength also changes. If the wave fronts are at a non-zero angle to the boundary, they change direction as well as changing speed. This effect is known as **refraction**.

Figure 10.15 shows the refraction of water waves in a ripple tank when they pass across a boundary, from deep to shallow water at a non-zero angle to the boundary. Because they move slower in the shallow water, the wavelength is smaller in the shallow water and therefore they change direction.

The direction of the refracted waves is closer to the **normal**, the line perpendicular to the boundary, than the direction of the incident waves. Refraction of light is observed when a light ray is directed into a glass block non-normally. The light ray changes direction when it crosses the glass boundary. This happens because light waves travel slower in glass than in air. (See section 11.1.)

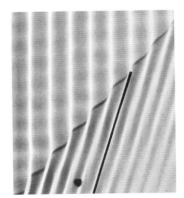

Origin of waves

Image point

Fig 10.14 *Reflection of circular waves*

Fig 10.15 *Refraction*

10.3.3 Diffraction

Diffraction occurs when waves spread out after passing through a gap, or round an obstacle. The effect can be seen in a ripple tank when straight waves are directed at a gap, as shown in Figure 10.16.

- The narrower the gap, the more the waves spread out.
- The longer the wavelength, the more the waves spread out.

To explain why the waves are diffracted on passing through the gap, consider each point on a wave front as a secondary emitter of 'wavelets'. The wavelets from the points along a wave front travel only in the direction in which the wave is travelling, not in the reverse direction, and they combine to form a new wave front spreading beyond the gap.

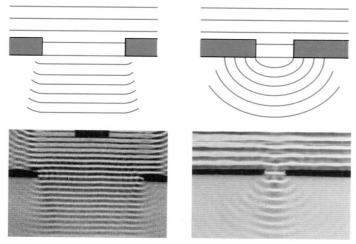

Fig 10.16 *The effect of the gap width*

QUESTIONS

1 Copy and complete Figure 10.17, by showing the wave front after it has been reflected from the straight reflector. Also, show the direction of the reflected wave front.

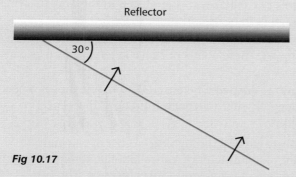

Fig 10.17

2 A circular wave spreads out from a point P on a water surface, which is 0.50 m from a flat reflecting wall. The wave travels at a speed of 0.20 m s⁻¹. Sketch the arrangement and show the position of the wavefront:

a 2.5 s,

b 4.0 s after the wave front was produced at P.

3 Copy and complete Figure 10.18, by showing the wave fronts after they have passed across the boundary and have been refracted. Also, show the direction of the refracted waves.

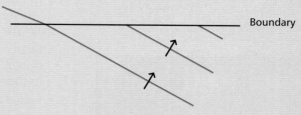

Fig 10.18

4 Water waves are diffracted on passing through a gap. How is the amount of diffraction changed as a result of:

a widening the gap without changing the wavelength,

b increasing the wavelength of the water waves without changing the gap width,

c increasing the wavelength of the water waves and reducing the gap width,

d widening the gap and increasing the wavelength of the waves?

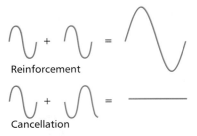

Reinforcement

Cancellation

Fig 10.19 *Superposition*

10.4.1 The Principle of Superposition

When waves meet, they pass through each other. At the point where they meet, they combine for an instant before they move apart. This combining effect is known as **superposition**. Imagine a boat hit by two wave crests at the same time from different directions. Anyone on the boat would know it had been hit by a **supercrest**, the combined effect of two wave crests.

> The Principle of Superposition states that when two waves meet, the total displacement at a point is equal to the sum of the individual displacements at that point.

- Where a crest meets a crest, a **supercrest** is created; the two waves reinforce each other.
- Where a trough meets a trough, a **supertrough** is created; the two waves reinforce each other.
- Where a crest meets a trough, the resultant displacement is **zero**; the two waves cancel each other out.

Further examples of superposition

1 Stationary waves on a rope

Stationary waves are formed on a rope if two people send waves continuously along a rope from either end, as shown in Figure 10.20. The two sets of waves are referred to as **progressive waves**, to distinguish them from stationary waves. They combine at fixed points along the rope to form points of no displacement, or **nodes**, along the rope. At each node, the two sets of waves are always 180° out of phase so they cancel each other out. Stationary waves are described in more detail in section 10.5.

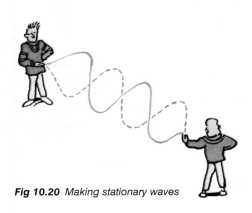

Fig 10.20 *Making stationary waves*

2 Water waves in a ripple tank

A vibrating dipper on a water surface sends out circular waves. Figure 10.21 shows a snapshot of two sets of circular waves produced in this way in a ripple tank. The waves pass through each other continuously:

- Points of **cancellation** are created where a crest from one dipper meets a trough from the other dipper. These points of cancellation are seen as gaps in the wave fronts.
- Points of **reinforcement** are created where a crest from one dipper meets a crest from the other dipper, or where a trough from one dipper meets a trough from the other dipper.

As the waves are continuously passing through each other at constant frequency and at a constant phase difference, cancellation and reinforcement occur at fixed positions. This effect is known as **interference**. The two dippers are said to be **coherent** emitters of waves, because they vibrate with a constant phase difference. If the phase difference changed at random, the points of cancellation and reinforcement would move about at random and no interference pattern would be seen. Interference of light is described in more detail in section 11.3.

Note The points of cancellation and reinforcement would be further apart if the wavelength was increased or the dipper were closer together.

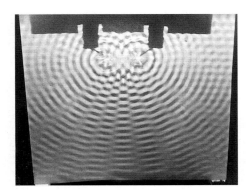

Fig 10.21 *Interference of water waves*

10.4.2 Tests using microwaves

A microwave transmitter and receiver can be used to demonstrate reflection, refraction, diffraction, interference and polarisation of microwaves. The transmitter produces microwaves of wavelength 3.0 cm. The receiver can be connected to a suitable meter, which gives a measure of the intensity of the microwaves at the receiver.

• Place the receiver in the path of the microwave beam from the transmitter. Move the receiver gradually away from the transmitter and note that the receiver signal decreases with distance from the transmitter. This shows that the microwaves become weaker as they travel away from the receiver.

• Place a metal plate between the transmitter and the receiver to show that microwaves cannot pass through metal.

• Use two metal plates to make a narrow slit and show that the receiver detects microwaves that have been diffracted as they pass through the slit. Show that if the slit is made wider, less diffraction occurs.

• Use a narrow metal plate with the two plates from above to make a pair of double slits, as in Figure 10.22. Direct the transmitter at the double slits and use the receiver to find points of cancellation and reinforcement where the microwaves from the two slits overlap.

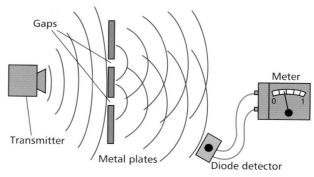

Fig 10.22 *Interference of microwaves*

QUESTIONS

1 Figure 10.23 shows two wave pulses on a rope travelling towards each other. Sketch a snapshot of the rope:

Fig 10.23

 a when the two waves are passing through each other,

 b when the two waves have passed through each other.

2 How would you expect the interference pattern in Figure 10.21 to change if:

 a the two dippers are moved further apart,

 b the frequency of the waves produced by the dippers is reduced?

3 Microwaves from a transmitter are directed at a narrow slit between two metal plates. A receiver is placed in the path of the diffracted microwaves, as shown in Figure 10.24.

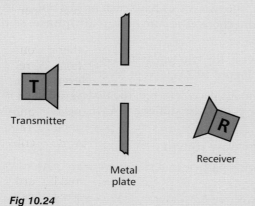

Fig 10.24

How would you expect the receiver signal to change if:

 a the receiver is moved directly away from the slit,

 b the slit is then made narrower?

4 Microwaves, from a transmitter, are directed at two parallel slits in a metal plate. A receiver is placed on the other side of the metal plate. When the receiver is moved a short distance along a line AB parallel to the plate, the receiver signal decreases then increases again.

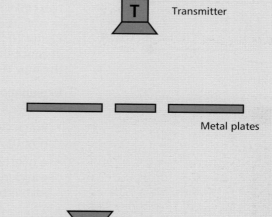

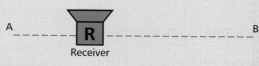

Fig 10.25

 a Explain why the signal decreased when it was moved from A along the line AB, and then increased.

 b Explain why the signal increased as it moved towards B.

Stationary and progressive waves

10.5.1 Formation of stationary waves

When a guitar string is plucked, the sound produced depends on the way in which the string vibrates. If the string is plucked gently at its centre, a stationary wave of constant frequency is set up on the string. The sound produced therefore has a constant frequency. If the guitar string is plucked harshly, the string vibrates in a more complicated way and the note produced contains other frequencies as well as the frequency produced when it is plucked gently.

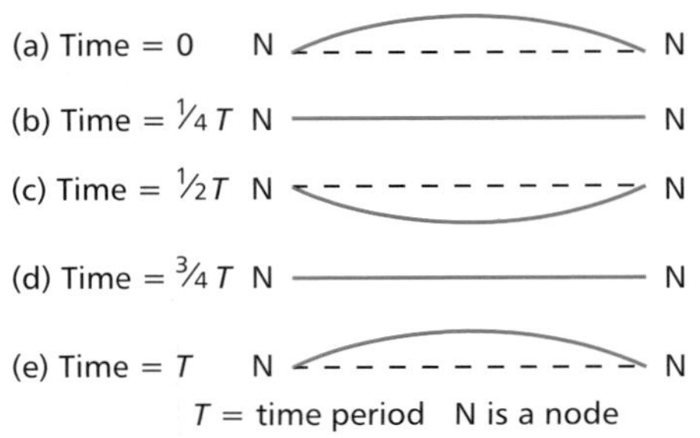

As explained in section 10.4, a stationary wave is formed when two progressive waves pass through each other. This can be achieved on a string in tension by:

- Sending progressive waves along the string from either end.

- Fixing one end of the string and sending progressive waves along it from the other end. The waves reflect at the fixed end and pass through progressive waves, moving towards the fixed end.

- Fixing both ends and making the middle part vibrate, so progressive waves travel towards each end, reflect at the ends, and then pass through each other.

The simplest stationary wave pattern on a string is shown in Figure 10.26. This is the **fundamental** mode of vibration of the string. It consists of a single loop that has a **node** (i.e. a point of no displacement) at either end. The string vibrates with maximum amplitude mid-way between the nodes. This position is referred to as an **antinode**. In effect, the string vibrates from side-to-side repeatedly. For this pattern to occur, the distance between the nodes at either end (i.e. the length of the string) must be equal to one half-wavelength of the waves on the string:

(a) Time = 0   N $\longleftarrow$ – – – – – – – $\longrightarrow$ N

(b) Time = $\frac{1}{4}T$ N $\underline{\hspace{4cm}}$ N

(c) Time = $\frac{1}{2}T$ N $\longleftarrow$ – – – – – – – $\longrightarrow$ N

(d) Time = $\frac{3}{4}T$ N $\underline{\hspace{4cm}}$ N

(e) Time = T N $\longleftarrow$ – – – – – – – $\longrightarrow$ N

T = time period   N is a node

Fig 10.26 *Fundamental vibrations*

Distance between adjacent nodes = $\frac{1}{2}\lambda$

If the frequency of the waves sent along the rope from either end is raised steadily, the pattern in Figure 10.26 disappears: a new pattern is observed with two equal loops along the rope. This pattern (Fig 10.27) has a node at the centre as well as at either end. It is formed when the frequency is twice as high as in Figure 10.26, corresponding to half the previous wavelength. Because the distance from one node to the next is equal to half a wavelength, the length of the rope is therefore equal to one full wavelength.

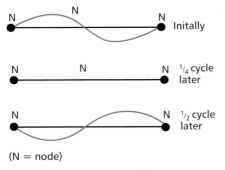

(N = node)

Fig 10.27 *A stationary wave of two loops*

Explanation of stationary waves

Consider a snapshot of two progressive waves passing through each other:

- When they are in phase, they reinforce each other to produce a large wave (Fig 10.28a).

- A quarter of a cycle later, the two waves have each moved one quarter of a wavelength in opposite directions. They are now in **antiphase**, so they cancel each other (Fig 10.28b).

- After a further quarter cycle, the two waves are back in phase. The resultant is again a large wave as in Figure 10.28a, except reversed.

The points where there is no displacement (i.e. the nodes) are fixed in position throughout. Between these points, the stationary wave oscillates between the nodes.

In general, in any stationary wave pattern:

The amplitude of a vibrating particle in a stationary wave pattern varies with position from:

- zero at a node,
- to maximum amplitude at an antinode.

Stationary waves do not transfer energy

The amplitude of vibration is zero at the nodes, so there is no energy at the nodes. The amplitude of vibration is a maximum at the antinodes, so there is maximum energy at the antinodes. Because the nodes and antinodes are at fixed positions, no energy is transferred in a stationary wave pattern.

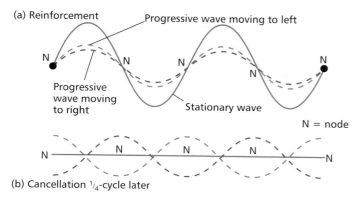

(a) Reinforcement

Progressive wave moving to left

Progressive wave moving to right

Stationary wave

N = node

(b) Cancellation ¼-cycle later

Fig 10.28 *Explaining stationary waves*

The phase difference between two vibrating particles:

- is zero if the two particles are between adjacent nodes or separated by an even number of nodes,
- is 180° if the two particles are separated by an odd number of nodes.

Table 10.2 *Comparison between stationary waves and progressive waves in terms of particle vibrations*

	Stationary waves	Progressive waves
Frequency	All particles, except at the nodes, vibrate at the same frequency.	All particles vibrate at the same frequency.
Amplitude	The amplitude varies from zero at the nodes to a maximum at the antinodes.	The amplitude is the same for all particles.
Phase difference between two particles	$m\pi$, where m is the number of nodes between the two particles.	$2\pi x/\lambda$, where x = distance apart and λ is the wavelength.

10.5.2 More examples of stationary waves

Sound in a pipe

Sound resonates at certain frequencies in an air-filled tube or pipe. In a pipe closed at one end, these resonant frequencies occur when there is an antinode at the open end and a node at the other end. (See section 10.7.)

Using microwaves

Microwaves from a transmitter are directed normally at a metal plate, which reflects the microwaves back towards the transmitter. When a detector is moved along the line between the transmitter and the metal plate, the detector signal is found to be zero at equally spaced positions along the line. The reflected waves and the waves from the transmitter form a stationary wave pattern. The positions where no signal is detected are where nodes occur. They are spaced at intervals of one half of a wavelength.

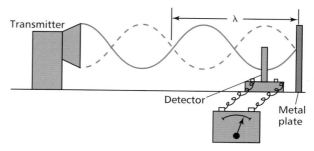

Transmitter

Detector

Metal plate

Fig 10.29 *Using microwaves*

QUESTIONS

1 **a** Sketch the stationary wave pattern seen on a rope, when there is a node at either end and an antinode at its centre.

   b If the rope in part **a** is 4.0 m in length, calculate the wavelength of the waves on the rope for the pattern you have drawn.

2 The stationary wave pattern shown in Figure 10.30 is set up on a rope of length 3.0 m.

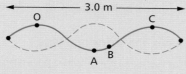

Fig 10.30

   a Calculate the wavelength of these waves.

   b State the phase difference between the particle vibrating at O and the particle vibrating at:
   (i) A
   (ii) B
   (iii) C

3 State two differences between a stationary wave and a progressive wave in terms of the vibrations of the particles.

4 Microwaves from a transmitter are reflected back to the transmitter using a flat metal plate. A detector is moved along the line between the transmitter and the metal plate. The detector signal is zero at positions 15 mm apart.

   a Explain why the signal is zero at certain positions.

   b Calculate the wavelength of the microwaves.

10.6.1 Stationary waves on a vibrating string

A controlled arrangement for producing stationary waves is shown in Figure 10.31. A string or wire is tied at one end to a mechanical vibrator, connected to a frequency generator. The other end of the string passes over a pulley and supports a weight, which keeps the tension in the string constant. As the frequency of the generator is increased from a very low value, different stationary wave patterns are seen on the string. In every case, the length of string between the pulley and the vibrator has a node at either end.

String at maximum displacement

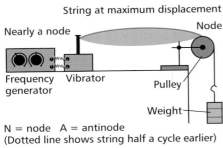

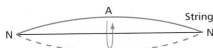

N = node   A = antinode
(Dotted line shows string half a cycle earlier)

(a) Fundamental

(b) First overtone

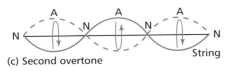

(c) Second overtone

Fig 10.31 *Stationary waves on a string*

- The **fundamental pattern** of vibration is seen at the lowest possible frequency. This has an antinode at the middle, as well as a node at either end. Because the length L of the vibrating section of the string is between adjacent nodes, the wavelength of the waves that form this pattern, the fundamental wavelength λ_0 is equal to $2L$. Therefore, the fundamental frequency:

$$f_0 = \frac{v}{2L}$$

where v is the speed of the progressive waves on the wire.

- The next stationary wave pattern, the **first overtone**, is where there is a node at the middle so the string is in two loops. The wavelength of the waves that form this pattern λ_1 is L, because each loop has a length of half a wavelength. Therefore, the frequency of the first overtone vibrations:

$$f_1 = \frac{v}{\lambda_1} = \frac{v}{L} = 2f_0$$

- The next stationary wave pattern, the **second overtone**, is where there are nodes at a distance of $\frac{1}{3}L$ from either end and an antinode at the middle. The wavelength of the waves that form this pattern λ_2 is $\frac{2}{3}L$, because each loop has a length of half a wavelength. Therefore, the frequency of the second overtone vibrations:

$$f_2 = \frac{v}{\lambda_2} = \frac{3v}{2L} = 3f_0$$

In general, stationary wave patterns occur at frequencies f_0, $2f_0$, $3f_0$, $4f_0$ etc., where f_0 is the frequency of the fundamental vibrations. This is the case in any vibrating linear system that has a node at either end.

Explanation of the stationary wave patterns on a vibrating string

In the arrangement shown in Figure 10.31, consider what happens to a wave sent out by the vibrator. The crest reverses its phase when it reflects at the fixed end, and travels back along the string as a trough. When it reaches the vibrator, it reflects and reverses phase again, travelling away from the vibrator once more as a crest. If this crest is reinforced by a crest created by the vibrator, the amplitude of the wave is increased. This is how a stationary wave is formed. The key condition is that the time taken for a wave to travel along the string and back should be equal to the time taken for a whole number of cycles of the vibrator:

- The time taken for a wave to travel along the string and back:

$$t = \frac{2L}{v}$$

where v is the speed of the waves on the string.

- The time taken for the vibrator to pass through a whole number of cycles:

$$t = \frac{m}{f}$$

where f is the frequency of the vibrator and m is a whole number (i.e. $m = 1$ or 2 or 3, etc.).

Therefore the key condition may be expressed as $\frac{2L}{\nu} = \frac{m}{f}$

Rearranging this equation gives:

$$f = \frac{m\nu}{2L} = mf_0 \text{ and } \lambda = \frac{\nu}{f} = \frac{2L}{m}$$

In other words:

- Stationary waves are formed at frequencies f_0, $2f_0$, $3f_0$, etc.
- The length of the vibrating section of the string

$L = \frac{m\lambda}{2}$ = whole number of half wavelengths

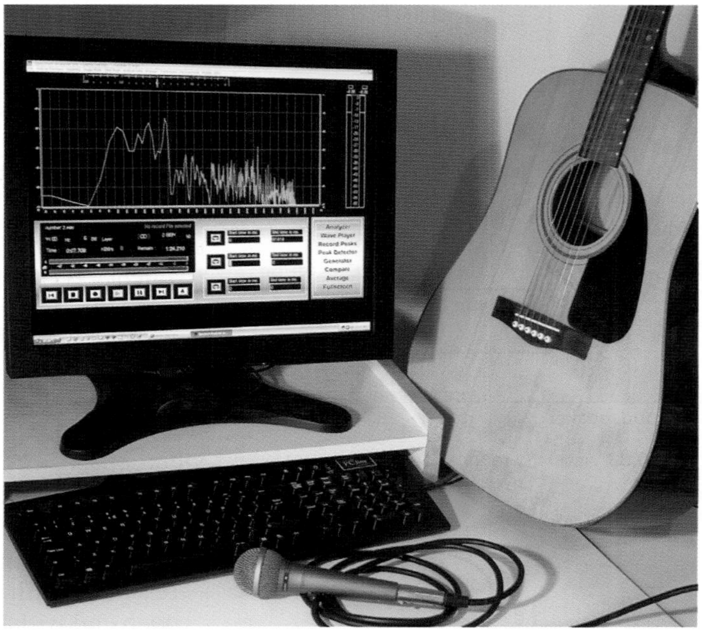

Fig 10.32 *A sound synthesiser*

10.6.2 Making music

A guitar produces sound when its strings vibrate as a result of being plucked. In an electronic guitar, the vibrations of the string are detected by a microphone which produces electrical waves. These are amplified and converted back to sound waves in a loudspeaker. In an acoustic guitar, the string vibrations make the guitar surfaces vibrate and send out sound waves.

When a stretched string or wire vibrates, its **pattern of vibration** is a mix of its fundamental mode of vibration and the overtones. The sound produced is the same mix of frequencies which change with time as the pattern of vibration changes. A **spectrum analyser** can be used to show how the intensity of a sound varies with frequency and with time. Combined with a **sound synthesiser**, the original sound can be altered by amplifying or suppressing different frequency ranges.

The **pitch** of a note produced by a stretched string can be altered by changing the **tension** of the string, or by altering its **length**. The pitch is raised by raising the tension or shortening the length. Lowering the tension or increasing the length lowers the pitch. By changing the length or altering the tension, a vibrating string or wire can be tuned to the same pitch as a vibrating tuning fork. However, the sound from a vibrating string includes the overtone frequencies as well as the fundamental frequency, whereas a tuning fork vibrates only at its fundamental frequency. The wire is tuned when its fundamental frequency is the same as the tuning fork frequency. A simple visual check is to balance a small piece of paper on the wire at its centre, and place the base of the vibrating tuning fork on one end of the wire. If the wire is tuned to the tuning fork, it will vibrate in its fundamental mode and the piece of paper will fall off.

QUESTIONS

1 A stretched wire of length 0.80 m vibrates in its fundamental mode at a frequency of 256 Hz. Calculate:

  a the wavelength of the progressive waves on the wire,

  b the speed of the progressive waves on the wire.

2 The fundamental frequency of vibration of a stretched wire is inversely proportional to the length of the wire. For the wire in question **1**, calculate the length of the wire needed to produce a frequency of:

  a 512 Hz

  b 384 Hz Assume that the tension of the wire is not changed.

3 Describe how you would expect the note from a vibrating guitar string to change if the string is:

  a shortened,

  b tightened.

4 The speed, ν, of the progressive waves on a stretched wire varies with the tension T in the wire, in accordance with the equation $\nu = \left(\frac{T}{\mu}\right)^{1/2}$, where μ (Greek *mu*) is the mass per unit length of the wire. Use this formula to explain why a nylon wire and a steel wire, of the same length, diameter and tension, produce notes of different pitch. State, with a reason, which wire would produce the higher pitch.

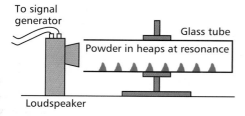

To signal generator

Glass tube

Powder in heaps at resonance

Loudspeaker

Fig 10.33 *Acoustic resonance*

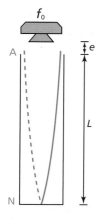

f_0

A

e

L

N

(a) Fundamental

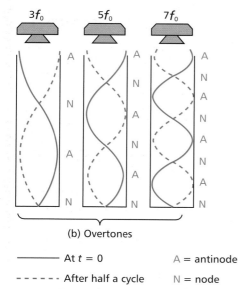

$3f_0$ $5f_0$ $7f_0$

(b) Overtones

——— At $t = 0$ A = antinode

- - - - - After half a cycle   N = node

Fig 10.34 *Resonances in a pipe closed at one end*

10.7.1 Stationary waves in a pipe closed at one end

If a small loudspeaker connected to a signal generator is placed near the open end of a pipe, the pipe resonates with sound at certain frequencies as the signal frequency is changed. Each resonance is due to stationary sound waves in the pipe. Sound waves from the transmitter at the open end are reflected at the other end of the pipe. The reflected waves, and the waves directly from the transmitter, pass along the pipe in opposite directions and therefore form a stationary wave pattern at certain frequencies. Figure 10.33 shows how this effect can be seen, as well as heard. The powder forms small heaps at the nodes, where there are no vibrations. Antinodes occur mid-way between the nodes and just beyond the open end. For greatest effect, the position of the loudspeaker needs to be at the antinode just beyond the open end.

- **At the fundamental frequency**, the lowest frequency at which the pipe resonates with sound, there is a node at the closed end and an antinode at the open end with no nodes or antinodes between. As shown in Figure 10.34, the pipe length $L = \frac{1}{4}\lambda_0$ (ignoring the small end-correction, e, at the open end). Therefore,

  the fundamental wavelength $\lambda_0 = 4L$,

  the fundamental frequency, $f_0 = \dfrac{v}{4L}$, where v is the speed of sound in the pipe.

- **Overtones** occur at frequencies $3f_0$, $5f_0$, $7f_0$, etc. In each case, there is a node at the closed end and an antinode at the open end, with one or more equally spaced nodes or antinodes between. As shown in Figure 10.34:

1 At the **first overtone**, the pipe length $L = \frac{3}{4}\lambda_1$ (ignoring the small end-correction at the open end). Therefore, the first overtone wavelength $\lambda_1 = \dfrac{4L}{3}$, where L is the pipe length, and

$$\text{the first overtone frequency, } f_1 = \frac{3v}{4L} = 3f_0$$

2 At the **second overtone**, the pipe length $L = \frac{5}{4}\lambda_2$ (ignoring the small end-correction).

   Therefore, the second overtone wavelength $\lambda_2 = \dfrac{4L}{5}$, where L is the pipe length,

   and the second overtone frequency, $f_2 = \dfrac{5v}{4L} = 5f_0$

Further resonances occur at $7f_0$, $9f_0$, etc., corresponding to an odd number of quarter wavelengths equal to the pipe length.

Note The above frequency formulas correspond to the condition that the time taken for sound to travel along the tube and back ($= \dfrac{2L}{v}$) is equal to the time for an odd number of half-cycles of vibration of the loudspeaker ($= (m + \frac{1}{2})/f$).

In this time a crest from the loudspeaker:
- travels the length of the tube,
- reflects and reverses phase at the closed end and returns as a trough to the open end,
- reflects, without reversal of phase, at the open end and is reinforced by a further trough from the loudspeaker.

10.7.2 Stationary waves in a pipe open at both ends

Sound waves travelling along the pipe partially reflect at the open end because the speed changes at the exit. An antinode is formed at either end. Therefore the pipe resonates with sound at any frequency corresponding to the pipe length L equal to a whole number of half wavelengths (the distance between two adjacent antinodes). Figure 10.35 shows the stationary wave patterns in this situation.

- **At the fundamental frequency,** there is an antinode at either end of the pipe with a node mid-way. As shown in Figure 10.35, the pipe length $L = \frac{1}{2}\lambda_0$ (ignoring the small end-corrections at each end). Therefore, the fundamental wavelength $\lambda_0 = 2L$, where L is the pipe length,

$$\text{the fundamental frequency, } f_0 = \frac{v}{2L},$$

where v is the speed of sound in the pipe.

- **Overtones** occur at frequencies $2f_0$, $3f_0$, $4f_0$, etc. In each case, there is an antinode at either end, with one or more equally spaced nodes and antinodes between. As shown in Figure 10.35,

1 At the **first overtone**, the wavelength $\lambda_1 =$ the pipe length L (ignoring the small end-corrections).

Therefore, the first overtone frequency, $f_1 = \dfrac{v}{L} = 2f_0$

2 At the **second overtone**, the wavelength λ_2 is such that the pipe length $L = \frac{3}{2}\lambda_2$ (ignoring the small end-corrections).

Therefore, the second overtone wavelength $\lambda_2 = \dfrac{2L}{3}$, where L is the pipe length,

the second overtone frequency, $f_2 = \dfrac{3v}{2L} = 3f_0$

Further resonances occur at $4f_0$, $5f_0$, etc., corresponding to a whole number of half-wavelengths equal to the pipe length.

Note The above frequency formulas correspond to the condition that the time taken for sound to travel the length of the tube and back $\left(= \dfrac{2L}{v}\right)$ is equal to the time for a whole number of cycles of vibration of the loudspeaker $\left(= \dfrac{m}{f}\right)$.

In this time, a crest from the loudspeaker:
- travels the length of the tube,
- reflects without phase reversal at the far end and returns as a crest to the loudspeaker end,
- reflects, without reversal of phase, at the loudspeaker end and is reinforced by a crest from the loudspeaker.

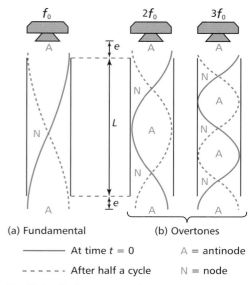

(a) Fundamental   (b) Overtones

——— At time $t = 0$ A = antinode

- - - - - After half a cycle   N = node

Fig 10.35 *Stationary waves in an open pipe*

QUESTIONS

1 A pipe, of length 1.20 m, is closed at one end and open at the other end. The speed of sound in the pipe is 340 m s^{-1}. Calculate:
   a its fundamental wavelength and frequency,
   b the frequency of its first overtone.

2 A pipe of length 1.50 m, closed at one end and open at the other end, resonates at a frequency of 170 Hz. The speed of sound in the pipe is 340 m s^{-1}. Calculate:
   a the wavelength of the sound waves in the pipe,
   b the fundamental frequency of this pipe.

3 A pipe, of length 2.40 m, is open at both ends. The speed of sound in the pipe is 340 m s^{-1}. Calculate:
   a its fundamental frequency,
   b its first overtone frequency.

4 A wind organ has pipes open at both ends, which are of lengths from 0.25 m to 2.50 m. Calculate the range of fundamental frequencies from these pipes. The speed of sound in the pipes is 340 m s^{-1}.

1 a (i) Explain the difference between longitudinal waves and transverse waves.
  (ii) State one example of a longitudinal wave.
  (iii) State one example of a transverse wave.

b With the aid of a diagram, explain what is meant by:
  (i) diffraction of a wave,
  (ii) refraction of a wave.

2 a (i) Explain what is meant by a 'polarised' transverse wave.
  (ii) Explain why sound waves cannot be polarised, whereas light waves can.

3 An oscilloscope is used to display the signal from a microphone when sound waves of constant frequency are directed at the microphone (see graph below).

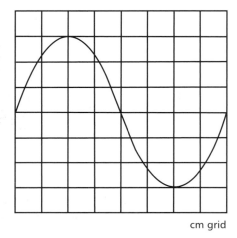

cm grid

a The oscilloscope time base is set at $0.5 \, \text{ms cm}^{-1}$. Calculate the frequency of the sound waves.

b Sketch the display on the oscilloscope you would observe, if the frequency of the sound waves was doubled with no change of loudness.

4 a Explain what is meant by the 'superposition' of waves.

b Two small loudspeakers, 1.2 m apart, are connected to a signal generator, as shown opposite. The signal generator is adjusted to produce an alternating voltage of constant amplitude, at a frequency of 1.0 kHz.
  (i) An observer moves along the line XY at a perpendicular distance of 2.0 m from the loudspeakers, as shown. The intensity of the sound rises and falls as she moves along the line. Explain this effect.
  (ii) In part (i), the observer notes that successive minima are further apart along the line XY when the test is repeated at a lower frequency. Explain this observation.

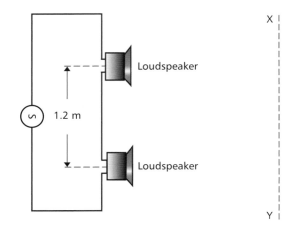

5 Microwaves from a transmitter were directed at two narrow gaps between three metal plates. A detector was placed on the other side of the plates at P, as shown below.

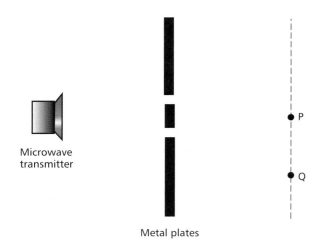

Metal plates

a The detector was moved along a line parallel to the plates to Q. The signal from the detector decreased from a maximum at P to a minimum mid-way between P and Q, then to a maximum at Q.
  (i) Explain why there was a maximum at P and at Q.
  (ii) Explain why there was a minimum mid-way between P and Q.

b Explain how the detector signal would change when the detector is mid-way between P and Q and a metal plate is placed over one of the gaps.

6 A stationary wave pattern of a wire, of length 0.60 m, vibrating at a frequency of 300 Hz is shown below.

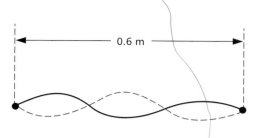

0.6 m

a In terms of amplitude and phase difference, compare the motion of the wire 0.10 m from the left-hand end with:
  (i) its motion at the mid-point,
  (ii) 0.10 m from the other end.

b (i) Calculate the wavelength of the waves on the wire.
  (ii) The frequency of vibrations is increased to 400 Hz, and a different stationary wave pattern is produced. Sketch the pattern you would expect to observe at 400 Hz.

7 Sound from a small loudspeaker connected to a signal generator was directed into a vertical glass tube containing some water, as shown below. The length of the air column in the tube could be varied by altering the level of water in the tube.

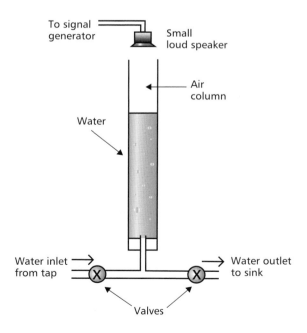

To signal generator
Small loud speaker
Air column
Water
Water inlet from tap
Water outlet to sink
Valves

a The length of the air column was gradually increased from a few centimetres, by lowering the level of water in the tube slowly. The tube resonated with sound at certain positions of the water level. Explain why this effect happened only at certain lengths of the air column.

b The tube resonated with sound of frequency 300 Hz when the length of the air column was 270 mm and 820 mm:
  (i) Diagram **a** shows how the amplitude of the sound waves in the tube varied with position along the tube when the air column length was 270 mm. Using a copy of diagram **b**, sketch the corresponding pattern when the air column length was 820 mm.
  (ii) Show that the wavelength of the sound in the tube was 1.10 m, and hence calculate the speed of sound in the tube.

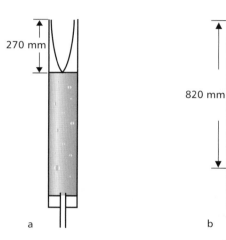

270 mm

820 mm

a   b

8 An open-ended pipe, of length 0.80 m, resonates with sound at a frequency of 200 Hz. The speed of sound in the pipe is 330 m s^{-1}.

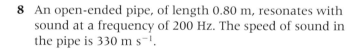

0.80 m

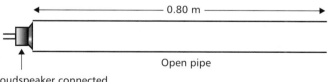

Open pipe

Loudspeaker connected to a signal generator

a (i) Calculate the wavelength of sound waves of frequency 200 Hz in the pipe.
  (ii) Sketch the stationary wave pattern in the pipe, when it resonates at 200 Hz.

b (i) Calculate the next highest frequency at which the pipe would resonate if the frequency was gradually increased.
  (ii) Explain why the pipe resonates at certain frequencies only.

Reflection and refraction of light

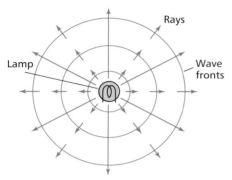

Fig 11.1 *Rays and waves*

The **wave theory of light** can be used to explain reflection and refraction of light. However, when we consider the effect of lenses or mirrors on the path of light, we usually prefer to draw diagrams using light rays. Light rays represent the direction of travel of wave fronts.

11.1.1 Reflection of light at a plane mirror

The laws of reflection

A **plane mirror** has a very smooth flat surface which reflects light waves without distorting or scattering the waves. Figure 11.2 shows the **incident ray**, representing the direction of the incident wave fronts, is at the same angle to the normal as the **reflected ray**, which represents the direction of the reflected wave fronts.

The two laws of reflection are:

1 The angle between the incident ray and the normal = the angle between the reflected ray and the normal.

2 The incident ray, the reflected ray and the normal are all in the same plane.

Image formation

When you observe the image of a point object in a plane mirror, you are looking at light that reflects off the mirror and appears to have come from a point behind the mirror. Figure 11.3 shows how an image is formed by a plane mirror. Every point of the image is due to light scattered from a corresponding point on the object. The image of a point object is the same distance behind the mirror as the object is in front. Consequently, the image of an extended object is **laterally inverted**.

The image is said to be **virtual**, because it is formed where the light appears to come from.

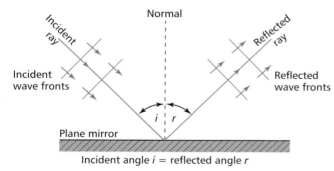

Fig 11.2 *Reflection by a plane mirror*

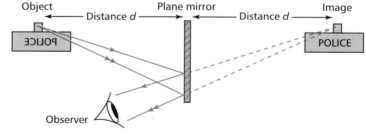

Fig 11.3 *Mirror images*

11.1.2 Refraction of light

Refraction is the change of direction that occurs when light passes non-normally across a boundary between two transparent substances. Figure 11.4 shows the change of direction of a light ray when it enters, and when it leaves, a rectangular glass block in air. The light ray bends:

- **towards** the normal when it passes from air into glass,
- **away from** the normal when it passes from glass into air.

No refraction takes place if the incident light ray is along the normal.

At a boundary between two transparent substances, the light ray bends:

- **towards** the normal if it passes into a **more** refractive substance,
- **away from** the normal if it passes into a **less** refractive substance.

Investigating the refraction of light by glass

Use a **ray box** to direct a light ray into a rectangular glass block at different angles of incidence at point P on one of the longer sides, as shown in Figure 11.4.

Note The **angle of incidence**, *i*, is the angle between the incident light ray and the normal at the point of incidence.

For each angle of incidence at P, mark the point Q where the light ray leaves the block. The angle of refraction is the angle between the normal at P and the line PQ. Measurements of the angles of incidence and refraction for different incident rays show that:

- The **angle of refraction**, *r*, at P is less than the angle of incidence, *i*,
- The ratio of $\dfrac{\sin i}{\sin r}$ is the same for each light ray.

  This is known as **Snell's Law**.

The ratio is referred to as the **refractive index, *n*,** of glass.

For a light ray travelling from air into a transparent substance:

Refractive index of the substance, $n = \dfrac{\sin i}{\sin r}$

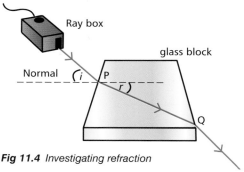

Fig 11.4 *Investigating refraction*

Worked example

A light ray is directed into a glass block of refractive index 1.5, at an angle of incidence of 40°. Calculate the angle of refraction of this light ray.

Solution

$i = 40°$, $n = 1.5$

Rearranging $\dfrac{\sin i}{\sin r} = n$ gives $\sin r = \dfrac{\sin i}{n}$

$$= \frac{\sin 40}{1.5} = \frac{0.643}{1.5} = 0.429$$

$$\therefore \qquad r = 25°$$

QUESTIONS

1 Figure 11.5 shows a light ray directed at a plane mirror at an angle of incidence of 60°. A second plane mirror is placed at right angles to the first mirror.

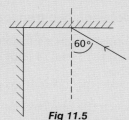

Fig 11.5

a Copy and complete the diagram, by showing the direction of the light ray after each reflection.

b Prove that the light ray leaves the second mirror in the opposite direction to the light ray incident on the first mirror.

2 A light ray is directed at a plane mirror at an angle of incidence of 35°:

a (i) Sketch the plane mirror, the incident ray, the reflected ray and the normal.

   (ii) Show that the reflected ray is at an angle of 70° to the incident ray.

b Show that if the mirror is turned through an angle of 5°, without changing the direction of the incident ray, the reflected ray moves through an angle of 10°.

3 A light ray is directed from air into a glass block of refractive index 1.5. Calculate:

a the angle of refraction at the point of incidence, if the angle of incidence is:
   (i) 30°   (ii) 60°

b the angle of incidence, if the angle of refraction at the point of incidence is:
   (i) 35°   (ii) 40°

4 A light ray in air was directed at the mid-point of the flat side of a semi-circular glass block at an angle of incidence of 50°, as shown in Figure 11.6. The angle of refraction at the point of incidence was 30°.

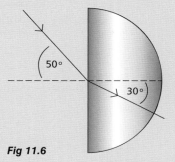

Fig 11.6

a Calculate the refractive index of the glass.

b Calculate the angle of refraction if the angle of incidence was changed to 60°.

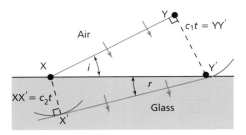

Fig 11.7 *Explaining refraction*

11.2.1 Explaining refraction

Refraction occurs because the **speed of the light waves** is different in each substance. The amount of refraction that takes place depends on the speed of the waves in each substance.

Consider a wave front of a wave when it passes across a straight boundary, as shown in Figure 11.7. Suppose the wave front moves from XY to X'Y' in time t, crossing the boundary between X and Y. In this time, the wave front moves:

• a distance $c_2 t$ at speed c_2 in the second substance from X to X', and
• a distance $c_1 t$ at speed c_1 in the first substance from Y to Y'.

Considering triangle XYY', since YY' is the direction of the wave front in substance 1 and is therefore perpendicular to XY, then:

$$YY' = XY' \sin i, \text{ where } i = \text{angle YXY'}$$
$$\therefore \qquad c_1 t = XY' \sin i$$

Considering triangle XX'Y', since XX' is the direction of the wave front in substance 2 and is therefore perpendicular to XY, then:

$$XX' = XY' \sin r, \text{ where } r = \text{angle XY'X'}$$
$$\therefore \qquad c_2 t = XY' \sin r$$

Combining these two equations gives:

$$\frac{\sin i}{\sin r} = \frac{c_1}{c_2}$$

This is in agreement with Snell's Law. If substance 1 is a vacuum, the ratio of the wave speeds $\dfrac{c_1}{c_2}$ is the **refractive index**, n, of the second substance.

11.2.2 Refraction at a boundary between two transparent substances

Consider a light ray which crosses a boundary from a substance in which the speed of light is c_1 to a substance in which the speed of light is c_2. As before:

$$\frac{\sin i}{\sin r} = \frac{c_1}{c_2}, \text{ where } i = \text{the angle between the incident ray and the normal,}$$
$$r = \text{the angle between the refracted ray and the normal.}$$

This equation may be rearranged as $\dfrac{1}{c_1} \sin i = \dfrac{1}{c_2} \sin r$

Multiplying both sides of this equation by c_0, the **speed of light in a vacuum**, therefore gives:

$$\frac{c_0}{c_1} \sin i = \frac{c_0}{c_2} \sin r$$

Substituting n_1 for $\dfrac{c_0}{c_1}$, where n_1 is the refractive index of substance 1

and n_2 for $\dfrac{c_0}{c_2}$, where n_2 is the refractive index of substance 2 gives:

$$n_1 \sin i = n_2 \sin r$$

Worked example

Water waves at a speed of 0.020 m s^{-1} cross a straight boundary and enter shallow water, where the speed of the waves is 0.015 m s^{-1}. Calculate the angle of refraction of the refracted waves when the angle of incidence is 30°.

Solution

Using $\dfrac{\sin i}{\sin r} = \dfrac{c_1}{c_2}$ with $c_1 = 0.020$ m s^{-1},

$c_2 = 0.015$ m s^{-1} and $i = 30°$

$$\frac{\sin 30}{\sin r} = \frac{0.020}{0.0150}$$

$$\therefore \qquad \sin r = \frac{0.015 \times \sin 30}{0.020} = 0.375$$

$$r = 22°$$

Worked example

A light ray crosses the boundary between water, of refractive index 1.3, and glass, of refractive index 1.5, at an angle of incidence of 40°. Calculate the angle of refraction of this light ray.

Solution

$n_1 = 1.3, n_2 = 1.5, i = 40°$

Using $n_1 \sin i = n_2 \sin r$ gives $1.3 \sin 40 = 1.5 \sin r$

$\therefore \qquad\qquad \sin r = \dfrac{1.3 \sin 40}{1.5} = 0.56$

$$r = 34°$$

Note The speed of light in air at atmospheric pressure is 99.97% of the speed of light in a vacuum. Therefore, the **refractive index of air** is 1.0003. For most purposes, the refractive index of air may be assumed to be 1.

Refraction of a light ray by a prism

Figure 11.9 shows the path of a light ray through a glass prism. The light ray refracts towards the normal where it enters the glass prism, then refracts away from the normal where it leaves the prism.

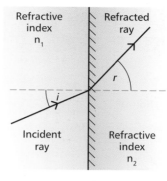

$$n_1 \sin i = n_2 \sin r$$

Fig 11.8 The 'n sin i' rule

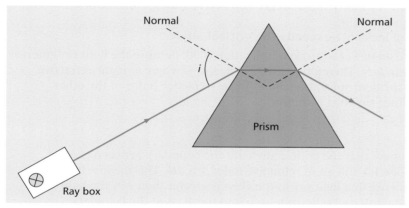

Fig 11.9 Refraction by a glass prism

11.2.3 Dispersion

Dispersion is the splitting of a beam of white light into the colours of the spectrum by a glass prism, as shown in Figure 11.10. This happens because white light is composed of light with a continuous range of wavelengths from **red**, at about 650 nm, to **violet**, at about 350 nm. The glass prism refracts light by different amounts, depending on its wavelength because the speed of light in glass depends on wavelength.

Fig 11.10 Dispersion of light

QUESTIONS

1 Water waves of frequency 4.0 Hz, travelling at a speed of 0.16 m s^{-1}, travel across a boundary from deep to shallow water where the speed is 0.12 m s^{-1}.

  a Calculate the wavelength of these waves:
    (i) in the deep water,
    (ii) in the shallow water.

  b The incident wave fronts cross the boundary at an angle of 25° to the boundary. Calculate the angle of the refracted wave fronts to the boundary.

2 **a** The speed of light in a vacuum is 3.00×10^8 m s^{-1}. Calculate the speed of light in:
    (i) glass of refractive index 1.52,
    (ii) water (of refractive index 1.33).

  b A light ray passes across a plane boundary from water into glass at an angle of incidence of 55°. Use the refractive index values from part **a** to show that the angle of refraction of this light ray is 46°.

3 Calculate the angle of refraction for a light ray entering glass, of refractive index 1.50, at an angle of incidence of 40° in:

  a air,

  b water (of refractive index 1.33).

4 A light ray enters an equilateral glass prism of refractive index 1.55 at the mid-point of one side of the prism at an angle of incidence of 35°:

  a Sketch this arrangement and show that the angle of refraction of the light ray in the glass is 22°.

  b (i) Show that the angle of incidence where the light ray leaves the glass prism is 38°.

    (ii) Calculate the angle of refraction of the light ray where it leaves the prism.

Total internal reflection

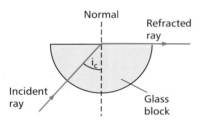

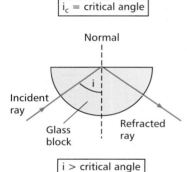

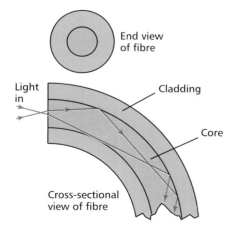

Fig 11.11 *Total internal reflection*

Fig 11.12 *An optical fibre*

11.3.1 Investigating total internal reflection

When a light ray travels from glass into air, it refracts away from the normal. If the angle of incidence is increased to a certain value known as the **critical angle,** the light ray refracts along the boundary. Figure 11.11 shows this effect. If the angle of incidence is increased further, the light ray undergoes **total internal reflection** at the boundary, the same as if the boundary was replaced by a plane mirror.

In general, total internal reflection can only take place if:

• the incident substance has a **larger refractive index** than the other substance,
• the angle of incidence **exceeds the critical angle**.

At the critical angle i_c, the angle of refraction is 90° because the light ray emerges along the boundary. Therefore, $n_1 \sin i_c = n_2 \sin 90$ where n_1 is the refractive index of the incident substance and n_2 is the refractive index of the other substance. Since $\sin 90 = 1$, then:

$$\sin i_c = \frac{n_2}{n_1}$$

Prove for yourself that the critical angle for the boundary between glass of refractive index 1.5 and air of refractive index 1 is 46°. This means that if the angle of incidence of a light ray in the glass is greater than 46°, the light ray undergoes total internal reflection back into the glass at the boundary.

11.3.2 Optical fibres

Optical fibres are used in medical endoscopes to see inside the body, and in communications to carry light signals. Figure 11.12 shows the path of a light ray along the core of an optical fibre. The light ray is totally internally reflected each time it reaches the core boundary.

• **A communications optical fibre** allows pulses of light that enter at one end from a transmitter to reach a receiver at the other end. Such fibres need to be highly transparent, to minimise absorption of light. Each fibre consists of a core surrounded by a layer of cladding of lower refractive index.

1 Total internal reflection takes place at the core–cladding boundary. At any point where two fibres are in direct contact, light would cross from one fibre to the other if there was no cladding. Such crossover would mean that the signals would not be secure, as they would reach the wrong destination.

2 The core must be very narrow to prevent **multi-path dispersion**. This occurs in a wide core, because light travelling along the axis of the core travels a shorter distance per metre of fibre than light that repeatedly undergoes total internal reflection. A pulse of light sent along a wide core would spread out and be longer than it ought to be. If it were too long, it would merge with the next pulse.

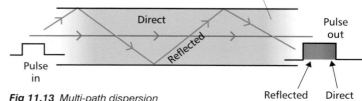

Fig 11.13 *Multi-path dispersion*

• **The medical endoscope** contains two bundles of fibres. The endoscope is inserted into a body cavity, which is then illuminated using light sent through one of the fibre bundles. A lens over the end of the other fibre bundle is used to form an image of the body cavity on the end of the fibre bundle. The light that forms this image travels along the fibres to the other end of the fibre bundle, where the image can be observed. This fibre bundle needs to be a **coherent** bundle, which means that the fibre ends at each end are in the same relative positions.

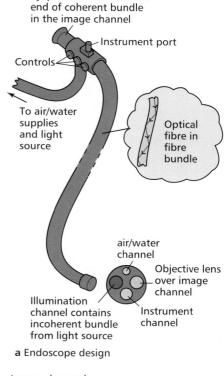

a Endoscope design

QUESTIONS

1 **a** State two conditions for a light ray to undergo total internal reflection at a boundary between two transparent substances.

b Calculate the critical angle for:
(i) glass of refractive index 1.52 and air,
(ii) water (of refractive index 1.33) and air.

2 **a** Show that the critical angle at a boundary between glass of refractive index 1.52 and water (of refractive index 1.33) is 61°.

b Figure 11.15 shows the path of a light ray in water (of refractive index 1.33) directed at an angle of incidence of 40° at a thick glass plate of refractive index 1.52. Calculate:
(i) the angle of refraction of the light ray at P,
(ii) the angle of incidence of the light ray at Q.

c Sketch the path of the light ray beyond Q.

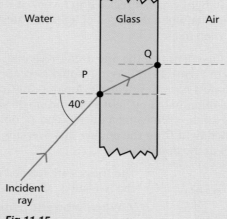

Fig 11.15

3 A window pane, made of glass of refractive index 1.55, is covered on one side only with a transparent film of refractive index 1.40.

a Calculate the critical angle of the film–glass boundary.

b A light ray in air is directed at the film at an angle of incidence of 45°, as shown in Figure 11.16. Calculate:
(i) the angle of refraction in the film,
(ii) the angle of refraction of the ray where it leaves the pane.

4 **a** In a medical endoscope, the fibre bundle used to view the image is coherent.
(i) What is meant by a 'coherent' fibre bundle?
(ii) Explain why this fibre bundle needs to be coherent.

b (i) Why is an optical fibre used in communication composed of a core surrounded by a layer of cladding of lower refractive index?
(ii) Why is it necessary for the core of an optical communications fibre to be narrow?

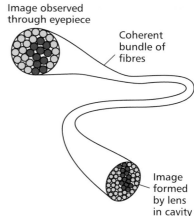

b A coherent bundle

Fig 11.14 The endoscope

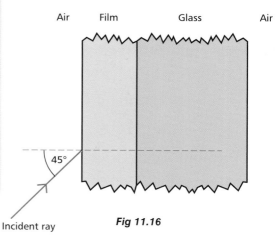

Fig 11.16

11.4 Interference of light

11.4.1 The wave nature of light

Young's double slits experiment

The wave nature of light was first suggested by Christian Huygens in the seventeenth century, but it was rejected at the time in favour of Sir Isaac Newton's corpuscular theory of light. Newton considered that light was composed of tiny particles he referred to as corpuscles and he was able to explain reflection and refraction using his theory. Huygens was also able to explain reflection and refraction using his wave theory. However, the two theories differed about whether or not light in a transparent substance travels faster (as predicted by Newton's theory) or slower (as predicted by Huygens' theory) than in air. Because of Newton's much stronger scientific reputation, Newton's theory of light remained unchallenged for over a century, until 1803, when Thomas Young at the Royal Institution first demonstrated interference of light. Even so, Newton's theory of light was not rejected in favour of Huygens' wave theory, until several decades later when the speed of light in water was measured and found to be slower than in air.

Fig 11.17 Thomas Young 1773–1829

An arrangement like the one used by Thomas Young to observe interference is shown in Figure 11.18. Young would have used a candle instead of a light bulb to illuminate a narrow single slit. A pair of narrow slits, referred to as the 'double slits', is illuminated by light from the single slit. Alternate bright and dark fringes can be seen using a microscope, or on a white screen placed where the diffracted light from the double slits overlaps. The fringes are evenly spaced and parallel to the double slits.

The fringes are formed due to **interference of light** from the two slits:

- Where a **bright fringe** is formed, the light from one slit reinforces the light from the other slit. In other words, the light waves from each slit arrive **in phase** with each other.
- Where a **dark fringe** is formed, the light from one slit cancels the light from the other slit. In other words, the light waves from the two slits arrive **180° out of phase**.

The distance from the centre of a bright fringe to the centre of the next bright fringe is called the **fringe separation, x**. This depends on the slit spacing a and the distance D from the slits to the screen, in accordance with the equation:

$$\textbf{Fringe separation, } x = \frac{\lambda D}{a},$$

where λ is the wavelength of light.

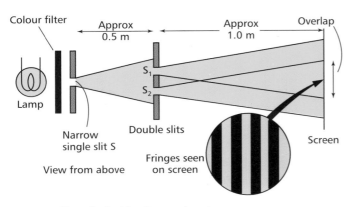

Fig 11.18 Young's double slits experiment

The equation shows that the fringes become more widely spaced if:

- the distance D from the slits to the screen is increased,
- the wavelength λ of the light used is increased,
- the slit spacing, a, is reduced.

Note The slit spacing is the distance between the **centres** of the slits.

11.4.2 The theory of the double slits equation

Consider the two slits S_1 and S_2 shown in Figure 11.19. At a point P on the screen where the fringes are observed, light emitted from S_1 arrives later than light from S_2 emitted at the same time. This is because the distance S_1P is greater than the distance S_2P. The difference between distances S_1P and S_2P is referred to as the **path difference**.

- **For reinforcement at P**, the path difference $S_1P - S_2P = m\lambda$, where $m = 0, 1, 2$, etc.
  Therefore, light emitted simultaneously from S_1 and S_2 arrives in phase at P, if reinforcement occurs at P.

- **For cancellation at P**, the path difference $S_1P - S_2P = (m + \frac{1}{2})\lambda$, where $m = 0, 1, 2$, etc.
  Therefore, light emitted simultaneously from S_1 and S_2 arrives at P out of phase by 180°, if cancellation occurs at P.

In Figure 11.19, a point Q along line S_1P has been marked such that $QP = S_2P$. Therefore, the path difference $S_1P - S_2P$ is represented by the distance S_1Q.

Because triangles S_1S_2Q and MOP are very nearly similar in shape, where M is the mid-point between the two slits and O is the mid-point of the central bright fringe of the pattern, then:

$$\frac{S_1Q}{S_1S_2} = \frac{OP}{OM}$$

If P is the mth bright fringe from the centre (where $m = 0, 1, 2$, etc.), then $S_1Q = m\lambda$ and $OP = mx$, where x is the distance between centres of adjacent bright fringes.

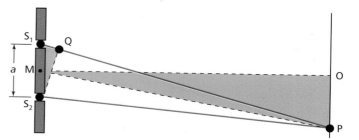

Fig 11.19 *The theory of the double slits experiment*

Also, OM = distance D and S_1S_2 = slit spacing a.

Therefore,

$$\frac{m\lambda}{a} = \frac{mx}{D}$$

Rearranging this equation gives: $\lambda = \dfrac{ax}{D}$

By measuring the slit spacing, a, the fringe separation, x, and the slit–screen distance D, the wavelength λ of the light used can be calculated. The formula is valid only if the fringe separation, a, is much less than the distance D from the slits to the screen. This condition is to ensure the triangles S_1S_2Q and MOP are very nearly similar in shape.

Note Light wavelengths are usually expressed in **nanometres (nm)**, where $1\text{ nm} = 10^{-9}$ m.

QUESTIONS

1 In a double slits experiment using red light, a fringe pattern is observed on a screen at a fixed distance from the double slits. How would the fringe pattern change if:

   a the screen was moved closer to the slits,

   b one of the double slits was blocked completely?

2 The following measurements were made in a double slits experiment:

   Slit spacing, $a = 0.4$ mm; fringe separation, $x = 1.1$ mm; slit–screen distance, $D = 0.80$ m.

   Calculate the wavelength of light used.

3 In question **2**, the double slits were replaced by a pair of slits with a slit spacing of 0.5 mm. Calculate the fringe separation for the same slit-screen distance and wavelength.

4 The following measurements were made in a double slits experiment:

   Slit spacing, $a = 0.4$ mm; fringe separation, $x = 1.1$ mm; wavelength of light used, $\lambda = 590$ nm.

   Calculate the distance from the slits to the screen.

11.5 More about interference

11.5.1 Coherence

The double slits are described as **coherent sources,** because they emit light waves with a constant phase difference. This is because each wave crest, or wave trough, from the single slit always passes through one of the double slits a fixed time after it passes through the other slit. The double slits therefore emit wave fronts with a constant phase difference.

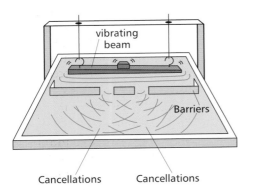

vibrating
beam

Barriers

Cancellations   Cancellations

Fig 11.20 *Interference of water waves*

The arrangement is like the ripple tank demonstration in Figure 11.20. Straight waves from the beam vibrating on the water surface diffract after passing through the two gaps in the barrier, and produce an interference pattern where the diffracted waves overlap. If one gap is closer to the beam than the other, each wave front from the beam passes through the nearer gap first. However, the time interval between the same wave front passing through the two gaps is always the same, so the waves emerge from the gaps with a constant phase difference.

Light from two separate light bulbs could not form an interference pattern, because the two light sources emit light waves at random. The points of cancellation and reinforcement would change at random, so no interference pattern is possible.

Wavelength and colour

In the double slits experiment, the fringe separation depends on the colour of light used. White light is composed of a continuous spectrum of colours, corresponding to a continuous range of wavelengths from about 350 nm for violet light to about 650 nm for red light. Each colour of light has its own wavelength, as shown in Figure 11.21.

The fringe patterns shown in Figure 11.22 show that the fringe separation is greater for red light than for blue light. This is because red light has a longer wavelength than blue light. The fringe spacing, x, depends on the wavelength, λ, of the light according to the formula: $\dfrac{x}{D} = \dfrac{\lambda}{a}$ as explained in section 11.4.

Rearranging this formula gives $x = \dfrac{\lambda D}{a}$. Thus the longer the wavelength of the light used, the greater the fringe separation is.

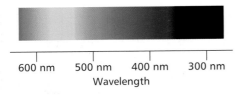

| 600 nm | 500 nm | 400 nm | 300 nm |

Wavelength

Fig 11.21 *Wavelength and colour*

Light sources

*   **Vapour lamps and discharge tubes** produce light with a dominant colour. For example, the sodium vapour lamp produces a yellow/orange glow which is due to light of wavelength 590 nm. Other wavelengths of light are also emitted from a sodium vapour lamp, but the colour due to light of wavelength 590 nm is much more intense than any other colour. A sodium vapour lamp is, in effect, a **monochromatic** light source because its spectrum is dominated by light of a certain colour.
*   **Light from a filament lamp or from the Sun** is composed of the colours of the spectrum, and therefore covers a continuous range of wavelengths from about 350 nm to about 650 nm.

Fig 11.22 *The double slits fringe pattern*

- **Light from a laser** is of a specific wavelength, and therefore laser light is highly monochromatic. For example, a helium–neon laser produces red light of wavelength 635 nm only. Because a laser beam is almost perfectly parallel and monochromatic, a convex lens can focus it to a very fine spot. The beam power is then concentrated in a very small area. This is why a laser beam is very dangerous if it enters the eye. The eye lens would focus the beam on a tiny spot on the retina and the intense concentration of light at that spot would destroy the retina.

> **Always wear safety goggles in the presence of a laser beam. Never look along a laser beam, even after reflection.**

11.5.2 Observing Young's fringes

Provided the light source is **not** a laser, a microscope may be used to observe the fringe pattern and to measure the fringe spacing. If so, the plane of viewing of the fringe pattern must be located to measure the distance D from the slits to the fringes accurately. Figure 11.23 shows how this can be done.

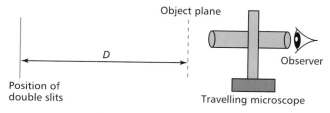

Fig 11.23 *Measuring the fringes using a microscope*

The contrast between the bright and dark fringes can be improved by narrowing the single slit (see Fig 11.18). If this slit is too wide, each part of it produces a fringe pattern which is displaced slightly from the pattern due to adjacent parts of the single slit. As a result, the dark fringes of the double slit pattern become narrower than the bright fringes, and contrast is lost between the dark and the bright fringes.

If a laser is used as the light source, a screen **must** be used on which to observe the fringe pattern produced by a laser. Never look along a laser beam, even after reflection. Safety goggles must always be worn in the presence of a laser beam.

White light fringes

Figure 11.22 shows the fringe patterns observed with blue light and with red light. As explained above, the blue light fringes are closer together than the red light fringes. The fringe pattern produced by white light is shown in Figure 11.24. Each component colour of white light produces its own fringe pattern, each pattern centred on the screen at the same position. As a result:

- The central fringe is **white**, because every colour contributes at the centre of the pattern.

- The inner fringes are tinged with **blue on the inner side and red on the outer side**. This is because the red fringes are more spaced out than the blue fringes, and the two fringe patterns do not overlap exactly.

- The outer fringes merge into an indistinct **background of white light**. This is because, where the fringes merge, different colours reinforce and therefore overlap.

Fig 11.24 *White light fringes*

QUESTIONS

1 a Sketch an arrangement that may be used to observe the fringe pattern observed when light from a narrow slit, illuminated by a sodium vapour lamp, is passed through a pair of double slits.

b Describe the fringe pattern you would expect to observe in part **a**.

2 In question **1**, describe how the fringe pattern would change if:

a one of the double slits is blocked,

b the narrow single slit is replaced by a wider slit.

3 Double slit interference fringes are observed using light of wavelength 590 nm and a pair of double slits of slit spacing 0.50 mm. The fringes are observed on a screen at a distance of 0.90 m from the double slits. Calculate the fringe separation of these fringes.

4 Describe and explain the fringe pattern that would be observed in question **3**, if the light source were replaced by a white light source.

Examination-style questions

1 a Choose from the list below to complete each of the following statements about how the frequency, speed and wavelength of light change when it passes from air into glass:

decreases   increases   stays the same

When light passes from air into glass:

(i) its speed _____

(ii) its frequency _____

(iii) its wavelength _____

b Light of wavelength 600 nm, travelling in air at a speed of 3.0×10^8 m s^{-1}, enters a liquid of refractive index 1.4. Calculate its speed, frequency and wavelength in the liquid.

2 The diagram shows a beam of red light directed from air into a rectangular glass block:

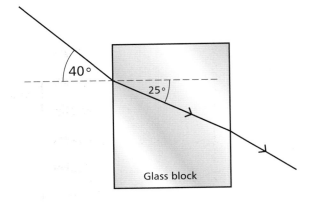

a The angle of refraction of the light ray at the point where the light ray entered the glass block was 25° when the angle of incidence was 40°.

(i) Calculate the refractive index of the glass for red light.

(ii) Calculate the angle of refraction if the angle of incidence had been 60°.

b White light consists of all the colours of the spectrum. The refractive index of glass is greater the smaller the wavelength of the incident light. Sketch what you would have observed if the beam of red light above had been replaced by a beam of white light at the same angle of incidence.

3 a (i) Explain, with the aid of a diagram, what is meant by 'total internal reflection'.

(ii) State two essential conditions for the total internal reflection of a light ray at a boundary between two transparent substances.

b The diagram shows a beam of yellow light directed normally into a semi-circular glass block through its curved surface. The refractive index of this glass block for yellow light is 1.5.

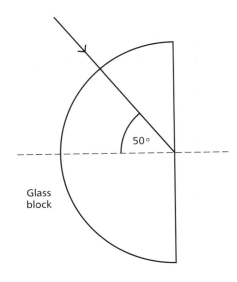

(i) Calculate the critical angle of yellow light in this glass block.

(ii) The light ray is incident on the flat surface at an angle of 50°. Sketch its path into and out of the glass block.

4 The refractive index of a transparent liquid can be measured by using two semi-circular glass blocks of known refractive index, pressed together with a film of the liquid in between, as shown below.

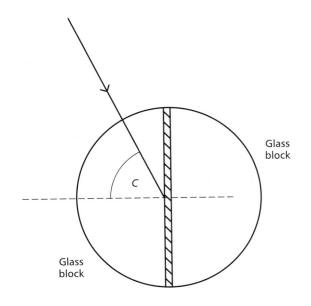

A narrow light ray is directed at the centre of the blocks through the curved face of one block. The angle of incidence, *i*, at the glass–liquid interface is increased until the light ray is totally internally reflected. At this position, the angle of incidence is equal to the critical angle, *c*, of the glass–liquid boundary.

a The method was tested using glass blocks of refractive index 1.48 and water (which has a refractive index of 1.33). Calculate the critical angle for the boundary between these two substances.

b Using a different liquid with the same glass blocks, the critical angle was 62°. Calculate the refractive index of the liquid.

5 Diagram **a** shows an optical fibre, which has a core of refractive index 1.49, surrounded by transparent cladding. The critical angle of the core–cladding boundary is 65°.

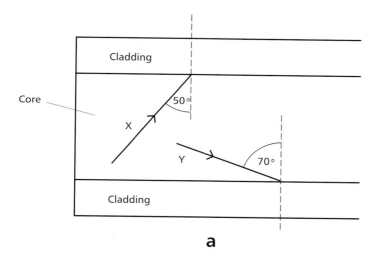

a

a (i) A light ray, X, in the core is incident on the boundary at an angle of incidence of 50°. Sketch the path of this light ray after the boundary.

(ii) A second light ray, Y, in the core is incident on the boundary at an angle of incidence of 70°. Sketch the path of this light ray after the boundary.

b A pulse of light enters the fibre at its flat end and travels a distance of 1000 m to the far end of the fibre.

(i) Calculate the least time taken for this pulse of light to travel along this length of fibre.

(ii) Show that the maximum time taken for this pulse of light to travel along this length of fibre is about 10% longer than the least time.

(iii) Diagram **b** shows how the intensity of the pulse of light varies with time, for the point where it enters the fibre. Sketch the corresponding intensity-time graph for the pulse when it reaches the far end of the fibre.

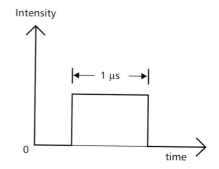

6 An interference pattern of bright and dark fringes is observed when two closely spaced slits are illuminated by a parallel beam of monochromatic light, as shown below:

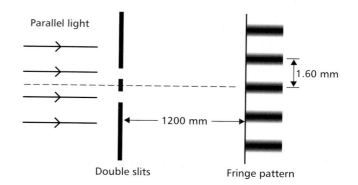

a Explain the formation of:
(i) a bright fringe,
(ii) a dark fringe.

b The two slits act as coherent emitters of light:
(i) Explain what is meant by this statement.
(ii) Explain why the fringe pattern could not be observed from two closely spaced light sources, such as two nearby filaments.

c The two slits were at a distance of 0.4 mm apart. The spacing between two adjacent bright fringes was 1.60 mm, when the fringes were observed at a distance of 1200 mm from the slits. Calculate the wavelength of the monochromatic light.

More on mathematical skills

Data handling

12.1.1 Scientific units

Scientists use a single system of units to avoid unnecessary effort and time converting between different units of the same quantity. This system, the **Système International d'Unités** (or SI system) is based on a defined unit for each of the five physical quantities listed in Table 12.1. Units of all other quantities are derived from these five SI base units.

Table 12.1 *SI base units*

Physical quantity	Unit
Mass	kilogram (kg)
Length	metre (m)
Time	second (s)
Electric current	ampere (A)
Temperature	kelvin (K)

The following examples show how the units of all other physical quantities are derived from the base units:

- The unit of **area** is the square metre (m^2).
- The unit of **volume** is the cubic metre (m^3).
- The unit of **density** is the kilogram per cubic metre ($kg\ m^{-3}$).
- The unit of **speed** is the metre per second ($m\ s^{-1}$).

12.1.2 More about using a calculator

Power of ten

Number displayed = 6.62×10^{-34}

Fig 12.1 *Displaying powers of ten*

EXP (or **EE** on some calculators)

This is the calculator button you press to key in a **power of ten**. To key in a number in standard form (e.g. 3.0×10^8), the steps are as follows:

Step 1 Key in the number between 1 and 10 (e.g. ③ . ⓪).

Step 2 Press the calculator button **EXP** (or **EE** on some calculators).

Step 3 Key in the power of ten (e.g. ⑧).

The display should now read *3.0   08* which should be read as 3.0×10^8 (not 3.0^8 which means 3.0 multiplied by itself 8 times). If the power of ten is a negative number (e.g. 10^{-8} not 10^8), press the calculator button [+/−] after Step 3 to change the sign of the power of ten.

Inv

This is the button you press if you want the calculator to give the value of the **inverse of a function**. For example, if you want to find out which angle has a sine of 0.5, you key in ⓪ . ⑤ , then **Inv** , then **sin** , to obtain the answer of 30°. Some calculators have a 'second function' button that you press instead of the 'inv' button.

log or **lg**

This is the button you press to find out what a number is as a power of ten. For example, press **log** , then key in ① ⓪ ⓪ and the display will show ② , because $100 = 10^2$. Logarithmic scales have equal intervals for each power of ten.

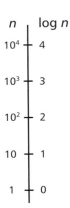

n	$\log n$
10^4	4
10^3	3
10^2	2
10	1
1	0

Fig 12.2 *A logarithmic scale*

y^x

This is used to raise any number to any power, for example, if you want to work out the value of 2^8, key in ② onto the display, then press the **y^x** button, then ⑧ and press = . The display should then show ② ⑤ ⑥ as the decimal value of 2^8.

The $\boxed{y^x}$ button can be used to find roots. For example, given the equation $T^4 = 5200$, you can find T by keying in $\boxed{5}\,\boxed{2}\,\boxed{0}\,\boxed{0}$ onto the display, then pressing the y^x button, followed by $(1 \div 4)$ which will give the answer 8.49.

Worked example

Calculate the cube root of 2.9×10^6.

Solution

Step 1 Key in 2.9×10^6 as explained earlier.

Step 2 Press the $\boxed{y^x}$ button.

Step 3 Key in $\boxed{(}\,\boxed{1}\,\boxed{\div}\,\boxed{3}\,\boxed{)}$.

Step 4 Press $\boxed{=}$.

The display should show $\boxed{1.426\ \ 02}$, so the answer is 142.6.

Significant figures

In general, always write a numerical answer to the same number of significant figures as the data used for the calculation. Because a calculator display shows a large number of digits, you should always round up, or round down, the answer displayed on a calculator to achieve the appropriate number of significant figures. For example, a calculator will show $\boxed{9.063077870 \times 10^{-1}}$ for the sine of $65°$. Because the data used ($65°$) is given to two significant figures, the sine of $65°$ should then be rounded off to 9.1×10^{-1}, and written as 0.91 because 9.06 to two significant figures is rounded up to 9.1.

QUESTIONS

Write your answers to each of the following questions in standard form where appropriate, and to the same number of significant figures as the data.

1 Copy and complete the following conversions:

a (i) 500 mm = … m
(ii) 3.2 m = … cm
(iii) 9560 cm = … m

b (i) 0.45 kg = … g
(ii) 1997 g = … kg
(iii) 54 000 kg = …g

c (i) 20 cm^2 = …m^2
(ii) 55 mm^2 = …m^2
(iii) 0.05 cm^2 = …m^2

2 **a** Write the following values in standard form:
(i) 150 million km in metres
(ii) 365 days in seconds
(iii) 630 nm in metres
(iv) 25.7 μg in kilograms
(v) 150 m in millimetres
(vi) 1.245 μm in metres

b Write the following values with a prefix instead of in standard form:
(i) 3.5×10^4 m = … km
(ii) 6.5×10^{-7} m = … nm
(iii) 3.4×10^6 g = …kg
(iv) 8.7×10^8 W = … MW = … GW

3 **a** Use the equation 'average speed = distance/time' to calculate the average speed in m s^{-1} of:
(i) a vehicle that travels a distance of 9000 m in 450 s,
(ii) a vehicle that travels a distance of 144 km in 2 h,
(iii) a particle that travels a distance of 0.30 nm in a time of 2.0×10^{-18} s,
(iv) the Earth on its orbit of radius 1.5×10^{11} m, given the time taken per orbit is 365.25 days.

b Use the equation 'resistance = $\dfrac{\text{potential difference}}{\text{current}}$', to calculate the resistance of a component for the following values of current I and p.d. V:
(i) $V = 15$ V, $I = 2.5$ mA
(ii) $V = 80$ mV, $I = 16$ mA
(iii) $V = 5.2$ kV, $I = 3.0$ mA
(iv) $V = 250$ V, $I = 0.51$ μA
(v) $V = 160$ mV, $I = 53$ mA

4 **a** Calculate each of the following:
(i) 6.7^3
(ii) $(5.3 \times 10^4)^2$
(iii) $(2.1 \times 10^{-6})^4$
(iv) $(0.035)^2$
(v) $(4.2 \times 10^8)^{1/2}$
(vi) $(3.8 \times 10^{-5})^{1/4}$

b Calculate each of the following:
(i) $\dfrac{2.4^2}{3.5 \times 10^3}$
(ii) $\dfrac{3.6 \times 10^{-3}}{6.2 \times 10^2}$
(iii) $\dfrac{8.1 \times 10^4 + 6.5 \times 10^3}{5.3 \times 10^4}$
(iv) $7.2 \times 10^{-3} + \dfrac{6.2 \times 10^4}{2.6 \times 10^6}$

12.2 Trigonometry

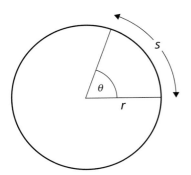

Fig 12.3 *Arcs and segments*

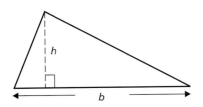

Area $= \frac{1}{2}hb$

a *The area of a triangle*

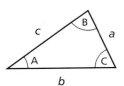

$$\frac{a}{\text{Sin A}} = \frac{a}{\text{Sin A}} = \frac{a}{\text{Sin A}}$$

Fig 12.4

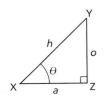

Fig 12.5 *A right-angled triangle*

12.2.1 More about the rules of trigonometry

Angles and arcs

- Angles are measured in **degrees or radians**. The scale for conversion is $360° = 2\pi$ radians. The symbol for the radian is rad so $1 \text{ rad} = \frac{360}{2\pi} = 57.3°$ (to three significant figures).

- The circumference of a circle of radius $r = 2\pi r$. So the circumference can be written as the angle in radians (2π) round the circle $\times r$.

- For a segment of a circle, the length of the arc of the segment is in proportion to the angle θ which the arc makes to the centre of the circle. This is shown in Figure 12.3. Because the arc length is $2\pi r$ (i.e. the circumference) for an angle of $360° (= 2\pi$ radians), then:

$$\frac{\text{Arc length, } s}{2\pi r} = \frac{\theta \text{ (in degrees)}}{360°}$$

More triangle rules

In addition to the 'right-angle triangle' rules described on p.7, the following rules apply to any triangle:

- Area of a triangle $= \frac{1}{2} \times$ its height $(h) \times$ its base (b)

- For a triangle with sides of lengths a, b and c, $\frac{a}{\sin A} = \frac{b}{\sin B} = \frac{c}{\sin C}$, where A, B and C are the angles opposite sides a, b and c. (See Figure 12.4.)

Trigonometry calculations using a calculator

A scientific calculator has a button you can press to use either degrees or radians. Make sure you know how to switch your calculator from one of these two modes to the other. Many marks have been lost in examinations as a result of forgetting to use the correct mode.

Trigonometry

Consider again the definitions of the sine, cosine and tangent of an angle, as applied to the right-angled triangle in Figure 12.5.

$$\sin \theta = \frac{o}{h} \qquad \text{where } o = \text{the length of the side opposite angle } \theta$$
$$h = \text{the length of the hypotenuse}$$
$$\cos \theta = \frac{a}{h} \qquad \text{and} \quad a = \text{the length of the side adjacent to angle } \theta$$
$$\tan \theta = \frac{o}{a}$$

Note $\tan \theta = \frac{o}{a} = \frac{o}{h} \times \frac{h}{a} = \frac{o}{h} \div \frac{a}{h} = \frac{\sin \theta}{\cos \theta}$

Pythagoras' theorem and trigonometry

Pythagoras' theorem states that for any right-angled triangle:

The square of the hypotenuse = the sum of the squares of the other two sides

Applying Pythagoras' theorem to the right-angled triangle in Figure 12.5 gives:

$$h^2 = o^2 + a^2$$

Since $o = h \sin \theta$ and $a = h \cos \theta$, then the above equation may be written:

$$h^2 = h^2 \sin^2 \theta + h^2 \cos^2 \theta$$

Cancelling h^2 therefore gives the following useful link between $\sin \theta$ and $\cos \theta$:

$$1 = \sin^2 \theta + \cos^2 \theta$$

12.2.2 Vector rules

Resolving a vector

As explained in section 1.1, any vector can be resolved into two perpendicular components in the same plane as the vector, as shown by Figure 12.6. The force vector F is resolved into a horizontal component $F \cos \theta$ and a vertical component $F \sin \theta$, where θ is the angle between the line of action of the force and the horizontal line.

Note The component at angle θ to the direction of the vector is always $\cos \theta \times$ the magnitude of the vector.

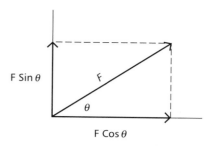

Fig 12.6 *Resolving force F*

Adding two perpendicular vectors

Figure 12.7 shows two perpendicular forces F_1 and F_2 acting on a point object X. The combined effect of these two forces, the resultant force, is given by the vector triangle. This is a right-angled triangle, where the resultant force is represented by the hypotenuse.

- Applying Pythagoras' theorem to the triangle gives $F^2 = F_1^2 + F_2^2$, where F is the magnitude of the resultant force. Therefore $F = (F_1^2 + F_2^2)^{1/2}$

- Applying the trigonometry formula $\tan \theta = \dfrac{o}{a}$ the angle between the resultant force and force F_1 is given by $\tan \theta = \dfrac{F_2}{F_1}$.

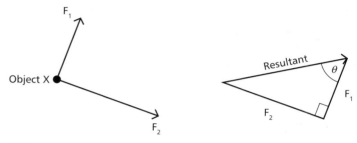

Fig 12.7 *Adding two perpendicular vectors*

QUESTIONS

1 a Calculate the circumference of a circle of radius 0.25 m.

b Calculate the length of the arc of a circle of radius 0.25 m for the following angles, between the arc and the centre of the circle:
   (i) 360°   (ii) 240°   (iii) 60°

2 For the right-angled triangle XYZ in Figure 12.8, calculate:

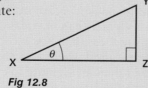

Fig 12.8

  a angle YXZ ($= \theta$),   **b** XZ if:
    if XY = 80 mm and: (i) XY = 20 cm and $\theta = 30°$
    (i) XZ = 30 mm   (ii) XY = 22 m and $\theta = 45°$
    (ii) XZ = 60 mm   (iii) YZ = 18 mm and $\theta = 75°$
    (iii) YZ = 30 mm   (iv) YZ = 47 cm and $\theta = 25°$
    (iv) YZ = 70 mm

3 a A right-angled triangle XYZ has a hypotenuse XY of length 55 mm and side XZ of length 25 mm. Calculate the length of the other side.

b An aircraft travels a distance of 30 km due North from an airport P to an airport Q. It then travels due East for a distance of 18 km, to an airport R.

Calculate:
  (i) the distance from P to R,
  (ii) the angle QPR.

4 a Calculate the perpendicular components A and B of each of the vectors in Figure 12.9.

b Calculate the magnitude and direction of the resultant of the two vectors shown in Figure 12.10.

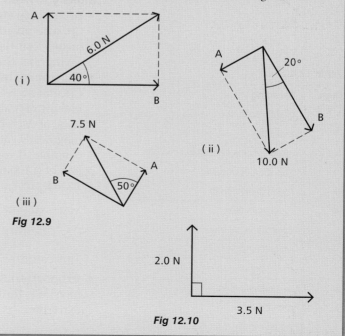

Fig 12.9

Fig 12.10

153

12.3.1 Signs and symbols

If you used symbols in your GCSE course, you might have met the use of s for distance and I for current. Maybe you wondered why we don't use d for distance instead of s; or C for current instead of I. The answer is that physics discoveries have taken place in many countries. The first person to discover the key ideas about speed was Galileo, the great Italian scientist; he used the word '*scale*' from his own language for distance, and therefore assigned the symbol s to distance. Important discoveries about electricity were made by Ampère, the great French scientist; he wrote about the intensity of an electric current, so he used the symbol I for electric current. The symbols we now use are used in all countries in association with the SI system of units.

Table 12.2 *Symbols for some physical quantities*

Physical quantity	Symbol	Unit	Unit symbol
Distance	s	metre	m
Speed or velocity	v	metre per second	m s^{-1}
Acceleration	a	metre per second per second	m s^{-2}
Mass	m	kilogram	kg
Force	F	newton	N
Energy or work	E	joule	J
Power	P	watt	W
Density	ρ	kilogram per cubic metre	kg m^{-3}
Current	I	ampere	A
Potential difference or voltage	V	volt	V
Resistance	R	ohm	

Signs you need to recognise

- **Inequality signs** are often used in physics. You need to be able to recognise the meaning of the signs in Table 12.3. For example, the inequality $I \geqslant 3$ A means that the current is greater than or equal to 3 A. This is the same as saying that the current is not less than 3 A.
- The **approximation sign** is used where an estimate or an order-of-magnitude calculation is made, rather than an accurate calculation. For an order-of-magnitude calculation, the final value is written with one significant figure only, or even rounded up or down to the nearest power of ten. Order-of-magnitude calculations are useful as a quick check after using a calculator. For example, if you are asked to calculate the density of a 1.0 kg metal cylinder of height 0.100 m and diameter 0.071 m, you ought to obtain a value of 2530 kg m^{-3} using a calculator. Now let's check the value:

$$\text{Volume} = \pi\,(\text{radius})^2 \times \text{height} \quad 3 \times (0.04)^2 \times 0.1 \quad 48 \times 10^{-5}\ \text{m}^3$$

$$\text{Density} = \frac{\text{mass}}{\text{volume}} \quad \frac{1.0}{50 \times 10^{-5}} \quad 2000\ \text{kg m}^{-3}$$

This confirms our 'accurate' calculation.

- **Proportionality** is represented by the $\propto$ sign. A simple example of its use in physics is for Hooke's Law: the tension in a spring is proportional to its extension.

$$\text{Tension},\ T \propto \text{extension},\ e$$

Table 12.3 *Signs*

Sign	Meaning
$>$	greater than
$<$	less than
$\geqslant$	greater than or equal to
$\leqslant$	less than or equal to
$\gg$	much greater than
$\ll$	much less than
	approximately equals
$<x>$	mean value
$<x^2>$	mean square value
$\propto$	is proportional to
Δ	change of
	square root

By introducing a constant of proportionality k, the link above can be made into an equation:

$$T = ke$$

where k is defined as the spring constant. See section 5.3.1. With any proportionality relationship, if one of the variables is increased by a given factor (e.g. $\times 3$), the other variable is increased by the same factor. So in the above example, if T is trebled, then extension e is also trebled. A graph of tension T on the y-axis against extension e on the x-axis would give a straight line through the origin.

12.3.2 More about equations and formulas

Rearranging an equation with several terms

The equation $v = u + at$ is an example of an equation with two terms on the right-hand side. These terms are u and at. To make t the subject of the equation:

- Isolate the term containing t on one side, by subtracting u from both sides, to give: $v - u = at$

- Isolate t by dividing both sides of the equation $v - u = at$ by a to give:

$$\frac{v - u}{a} = \frac{at}{a} = t$$

Note a cancels out in the expression $\frac{at}{a}$.

- The rearranged equation may now be written: $t = \frac{v - u}{a}$

Rearranging an equation containing powers

Suppose a quantity is raised to a power in a term in an equation and that quantity is to be made the subject of the equation. For example, consider the equation $V = \frac{4}{3}\pi r^3$ where r is to be made the subject of the equation:

- Isolate r^3 from the other factors in the equation by dividing both sides by 4π then multiplying both sides by 3 to give: $\frac{3V}{4\pi} = r^3$

- Take the cube root of both sides to give: $\left(\frac{3V}{4\pi}\right)^{1/3} = r$

- Rewrite the equation with r on the left-hand side if necessary.

More about powers

- Powers add for identical quantities when two terms are multiplied together. For example, if $y = ax^m$ and $z = bx^n$, then: $yz = ax^m bx^n = abx^{m + n}$

- An equation of the form $y = \frac{k}{z^n}$ may be written in the form: $y = kz^{-n}$

- The nth root of an expression is written as the power $\frac{1}{n}$. For example, the square root of x is $x^{1/2}$. Therefore, rearranging $y = x^n$ to make x the subject gives:

$$x = y^{1/n}$$

QUESTIONS

1 Complete each of the following statements:

a If $x > 5$, then $\frac{1}{x} < \ldots$

b If $4 < x < 10$, then $\ldots < \frac{1}{x} < \ldots$

c If $x^2 > 100$ then $\frac{1}{x} \ldots$

2 a Make t the subject of each of the following equations:
   (i) $v = u + at$ (ii) $s = \frac{1}{2}at^2$

   (iii) $y = k\,(t - t_0)$ (iv) $F = \frac{mv}{t}$

   b Solve each of the following equations:
   (i) $2z + 6 = 10$ (ii) $2\,(z + 6) = 10$
   (iii) $\frac{2}{z - 4} = 8$ (iv) $\frac{4}{z^2} = 36$

3 a Make x the subject of each of the following equations:
   (i) $y = 2x^{1/2}$ (ii) $2y = x^{-1/2}$

   (iii) $yx^{1/3} = 1$ (iv) $y = \frac{k}{x^2}$

b Solve each of the following equations:
   (i) $x^{-1/2} = 2$

   (ii) $3x^2 = 24$

   (iii) $\frac{8}{x^2} = 32$

   (iv) $2(x^{1/2} + 4) = 12$

4 Use the data given with each equation below to calculate:

   a The area of cross-section A, of a wire of radius $r = 0.34$ mm and length $L = 0.840$ m, using the equation $A = \pi r^2 L$.

   b The radius r of a sphere, of volume $V = 1.00 \times 10^{-6}$ m³, using the formula $V = \frac{4}{3}\pi r^3$.

   c The time period T of a simple pendulum of length $L = 1.50$ m, using the formula $T = 2\pi(L/g)^{0.5}$, where $g = 9.8$ m s⁻².

   d The speed v of an object, of mass $m = 0.20$ kg and kinetic energy $E_k = 28$ J, using the formula

$$E_k = \frac{1}{2}mv^2.$$

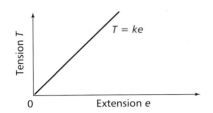

Fig 12.11

12.4.1 The general equation for a straight-line graph

Links between two physical quantities can be established most easily by plotting a graph. One of the physical quantities is represented by the vertical scale (the 'ordinate', often called the y-axis) and the other quantity by the horizontal scale (the 'abscissa', often called the x-axis). The **coordinates** of a point on a graph are the x- and y-values, usually written (x, y) of the point.

The simplest link between two physical variables is where the plotted points define a straight line. For example, Figure 12.11 shows the link between the tension in a spring and the extension of the spring; the gradient of the line is constant and the line passes through the origin. Any situation where the y-variable is proportional to the x-variable gives a straight line through the origin. For Figure 12.11, the gradient of the line is the spring contant k. The relationship between the tension T and the extension e may therefore be written as $T = ke$.

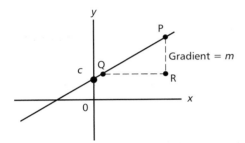

Fig 12.12 $y = mx + c$

The general equation for a straight-line graph is usually written in the form:

$$y = mx + c$$

where $m =$ the gradient of the line, and $c =$ the y-intercept.

- The **gradient m** can be measured by marking two points, P and Q, as far apart as possible on the line. The triangle PQR as shown in Figure 12.12 is then used to find the gradient. If (x_P, y_P) and (x_Q, y_Q) represent the x- and y-coordinates of points P and Q respectively, then

$$\text{gradient } m = \frac{y_P - y_Q}{x_P - x_Q}$$

- The **y-intercept, c**, is the point at $x = 0$, where the line crosses the y-axis. To find the y-intercept of a line on a graph that does not show $x = 0$, measure the gradient as above then use the coordinates of any point on the line with the equation $y = mx + c$ to calculate c. For example, rearranging $y = mx + c$ gives $c = y - mx$. Therefore, using the coordinates of point Q in Figure 12.12, the y-intercept $c = y_Q - mx_Q$.

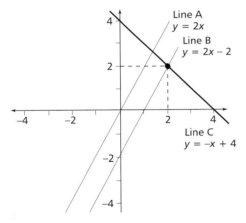

Fig 12.13 *Straight-line graphs*

Examples of straight-line graphs

- **Line A**: $c = 0$, so the line passes through the origin; its equation is $y = 2x$.
- **Line B**: $m > 0$, so the line has a positive gradient; its equation is $y = 2x - 2$.
- **Line C**: $m < 0$, so the line has a negative gradient; its equation is $y = -x + 4$.

12.4.2 Straight-line graphs and physics equations

You need to be able to be able to work out gradients and intercepts for equations you meet in physics that generate straight-line graphs. Some further examples are described below:

Motion at constant acceleration

The velocity v of an object moving at constant acceleration a at time t is given by the equation $v = u + at$, where u is its speed at time t. Figure 12.14 shows the corresponding graph of velocity v on the y-axis against time t on the x-axis.

Rearranging the equation as $v = at + u$ and comparing this with $y = mx + c$ shows that:

- the gradient, $m =$ acceleration a, and
- the y-intercept, $c =$ the initial velocity u.

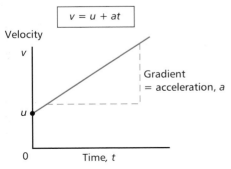

Fig 12.14 *Motion at constant acceleration*

P.d. and current for a battery

The p.d. V across the terminals of a battery, of e.m.f. ε and internal resistance r, varies with current in accordance with the equation $V = \varepsilon - Ir$ (see section 7.3). Figure 12.15 shows the corresponding graph of p.d. V on the y-axis against current I on the x-axis.

Rearranging the equation as $V = -rI + \varepsilon$ and comparing this with $y = mx + c$ shows that:

- the gradient, $m = -r$, and
- the y-intercept, $c = \varepsilon$ so the intercept on the y-axis gives the e.m.f. ε of the battery.

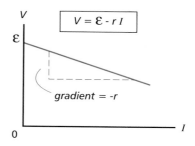

Fig 12.15 *P.d v. current for a battery*

Photoelectric emission

The maximum kinetic energy E_{Kmax} of a photoelectron, emitted from a metal surface of work function ϕ, varies with frequency f of the incident radiation in accordance with the equation $E_{Kmax} = hf - \phi$ (see section 9.4). Figure 12.16 shows the corresponding graph of E_{Kmax} on the y-axis against f on the x-axis.

Comparing the equation $E_{Kmax} = hf - \phi$ with $y = mx + c$ shows that:

- the gradient, $m = h$, and
- the y-intercept, $c = -\phi$.

Note The x-intercept is where $y = 0$ on the line. Let the coordinates of the x-intercept be $(x_0, 0)$. Therefore $mx_0 + c = 0$ so $x_0 = \dfrac{-c}{m}$. In Figure 12.16, the x-intercept is therefore $\dfrac{\phi}{h}$. This is the threshold frequency f_0, $= \dfrac{\phi}{h}$.

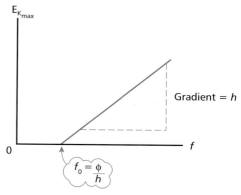

Fig 12.16 *Photoelectric emission*

Simultaneous equations

In physics, simultaneous equations can be solved graphically by plotting the line for each equation. The solution of the equations is given by the coordinates of the points where the lines meet. For example, lines B and C in Figure 12.12 meet at the point (2, 2) so $x = 2$, $y = 2$ are the only values of x and y that fit both equations.

Solving simultaneous equations doesn't require graph plotting if the equations can be arranged to fit one of the variables. Start by rearranging to make y the subject of each equation if necessary. Considering the example above:

- **Line B**; $y = 2x - 2$
- **Line C**; $y = -x + 4$

At the point where they meet, their coordinates are the same so solving $2x - 2 = -x + 4$ gives $x = 2$. Since $y = 2x - 2$, then $y = (2 \times 2) - 2 = 2$.

QUESTIONS

1 For each of the following equations that represent straight-line graphs, write down:
  (i) the gradient,
  (ii) the y-intercept,
  (iii) the x-intercept:

  a $y = 3x - 3$ **b** $y = -4x + 8$
  c $y + x = 5$ **d** $2y + 3x = 6$

2 **a** A straight line on a graph has a gradient $m = 2$ and passes through the point $(2, -4)$. Work out:
  (i) the equation for this line,
  (ii) its y-intercept.

  b The velocity v (in m s^{-1}) of an object varies with time t (in s) in accordance with the equation $v = 5 + 3t$.

Determine:
  (i) the acceleration of the object,
  (ii) the initial velocity of the object.

3 **a** Plot the equations $y = x + 3$ and $y = -2x + 6$ over the range from $x = -3$ to $x = +3$. Write down the coordinates of the point P where the two lines cross.

  b Write down the equation for the line OP, where O is the origin of the graph.

4 Solve the following pairs of simultaneous equations, after making y the subject of each equation if necessary:
  a $y = 2x - 4$, $y = -x + 2$
  b $y = 3x - 4$, $x + y = 8$
  c $2x + 3y = 4$, $x + 2y = 2$

More on graphs

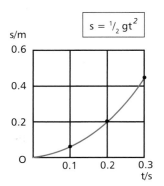

Fig 12.17 *s* v. *t*

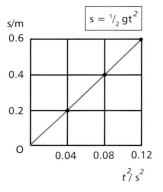

Fig 12.18 *s* v. *t²*

12.5.1 Curves and equations

Graphs with curves occur in physics in two ways:

- In practical work, where one physical variable is plotted against another and the plotted points do not lie along a straight line. For example, a graph of p.d. on the *y*-axis against current on the *x*-axis for a filament lamp is a curve that passes through the origin.

- In theory work, where an equation containing two physical variables is not in the form of the equation for a straight line ($y = mx + c$). For example, for an object released from rest, a graph of distance fallen, *s*, on the *y*-axis against time, *t*, on the *x*-axis is a curve because $s = \frac{1}{2}gt^2$. Figure 12.17 shows this equation represented on a graph.

Knowledge of the general equations for some common curves is an essential part of physics. When a curve is produced as a result of plotting a set of measurements in a practical experiment, few conclusions can be drawn as the exact shape of a curve is difficult to test. In comparison, if the measurements produce a straight line, it can then be concluded that the two physical variables plotted on the axes are related by an equation of the form $y = mx + c$.

If a set of measurements produces a curve rather than a straight line, knowledge of the theory could enable the measurements to be processed in order to give a straight-line graph which would then be a confirmation of the theory. For example, the distance and time measurements that produced the curve in Figure 12.17 could be plotted as distance fallen, *s*, on the *y*-axis against t^2 on the *x*-axis (where *t* is the time taken). Figure 12.18 shows the idea.

If a graph of *s* against t^2 gives a straight line, this would confirm that the relationship between *s* and *t* is of the form $s = kt^2$, where *k* is a constant. Because theory gives $s = \frac{1}{2}gt^2$, it can then be concluded that the theory applies to this set of measurements and that $k = \frac{1}{2}g$.

12.5.2 From curves to straight lines

Parabolic curves

Parabolic curves describe the flight paths of projectiles, or other objects, acted on by a constant force that is not in the same direction as the initial velocity of the object. In addition, parabolic curves occur where the energy of an object depends on some physical variable.

The general equation for a parabola is $y = kx^2$. Figure 12.19 shows the shape of the parabola $y = 3x^2$. Equations of the form $y = kx^2$ pass through the origin and they are symmetrical about the *y*-axis. This is because equal positive and negative values of *x* always give the same *y*-value.

- **The flight path for a projectile** projected horizontally at speed *u* has coordinates:

$$x = ut, \quad y = \tfrac{1}{2}gt^2$$

where *x* = horizontal distance travelled, *y* = vertical distance fallen and *t* is the time from initial projection (see section 1.8).

Combining these equations gives the flight path equation:

$$y = \frac{gx^2}{2u^2}$$

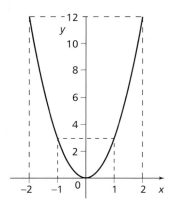

Fig 12.19 $y = 3x^2$

which is the same as the parabola equation $y = kx^2$, where $\frac{g}{2u^2}$ is represented by k in the equation.

A set of measurements plotted as a graph of vertical distance fallen, y against horizontal distance travelled, x, would be a parabolic curve as shown in Figure 12.19. However, a graph of y against x^2 should give a straight line (of gradient k) through the origin because $y = kx^2$.

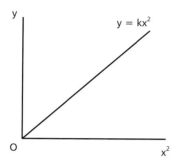

Fig 12.20 y against x^2

Inverse curves

An **inverse** relationship between two variables x and y is of the form $y = \frac{k}{x}$, where k is a constant. The variable y is said to be **inversely proportional** to variable x. If x is doubled, y is halved. If x is increased tenfold, y decreases to a tenth.

Figure 12.21 shows the curve for $y = \frac{10}{x}$.

The curve tends towards either axis, but never actually meets the axes. The correct mathematical word for 'tending towards but never meeting' is **asymptotic**. Consider the following example.

- The resistance R, of a wire of constant length L, varies with the wire's area of cross-section A in accordance with the equation:

$$R = \frac{\rho L}{A}$$

where ρ is the resistivity of the wire. See section 6.3. Therefore, R is inversely proportional to A. A graph of R (on the vertical axis) against A would therefore be a curve like Fig 12.21. However, a graph of R (on the vertical axis) against $\frac{1}{A}$ would be a straight line through the origin. The gradient of this straight line would be ρL.

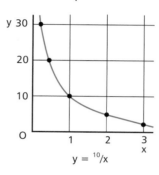

$y = {}^{10}/x$

Fig 12.21 $y = \frac{10}{x}$

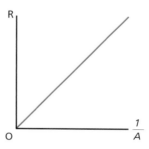

R against $\frac{1}{A}$

Fig 12.22 The resistance of a wire

QUESTIONS

1 The potential energy, E_P, stored in a stretched spring varies with the extension, e, of the spring in accordance with the equation $E_P = \frac{1}{2}ke^2$. Sketch a graph of E_P against:

   a e **b** e^2

2 The energy E_{ph} of a photon varies with its wavelength λ in accordance with the equation $E_{ph} = \frac{hc}{\lambda}$, where h is the Planck constant and c is the speed of light. Sketch a graph of E_{ph} against:

   a λ **b** $\frac{1}{\lambda}$

3 The current I through a wire of resistivity ρ varies with the length L, area of cross-section A and p.d. V in accordance with the equation $I = \frac{VA}{\rho L}$.

a Sketch a graph of I against: (i) V (ii) L (iii) $\frac{1}{L}$

b Explain how you would determine the resistivity from the graph of I against $\frac{1}{L}$.

4 An object released from rest falls at constant acceleration a and passes through a horizontal light beam at speed u. The distance it falls in time t after passing through the light beam is given by the equation $s = ut + \frac{1}{2}at^2$.

a Show that: $\frac{s}{t} = u + \frac{1}{2}at$

b (i) Sketch a graph of $\frac{s}{t}$ on the vertical axis against t on the horizontal axis.

    (ii) Explain how u and a can be determined from the graph.

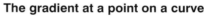

12.6 Graphs, gradients and areas

12.6.1 Gradients

The gradient of a straight line

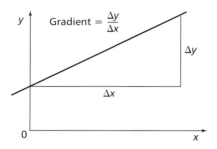

Fig 12.23 Constant gradient

The gradient of a straight line $= \frac{\Delta y}{\Delta x}$, where Δy is the change of the quantity plotted on the y-axis and Δx is the change of the quantity plotted on the x-axis. The gradient of a straight line is obtained by drawing as large a gradient triangle as possible and measuring the height Δy and the base Δx of this triangle, using the scale on each axis (Fig 12.23).

Note As a rule, when you plot a straight-line graph, always choose a scale for each axis that covers at least half the length of each axis. This will enable you to draw the line of best fit as accurately as possible, as explained in section 13.5.2. The measurement of the gradient of the line will therefore be more accurate. If the y-intercept is required and it cannot be read directly from the graph, it can be calculated by substituting the value of the gradient and the coordinates of a point on the line into the equation $y = mx + c$.

The gradient at a point on a curve

The gradient of a point on a curve is equal to the gradient of the tangent to the curve at that point.

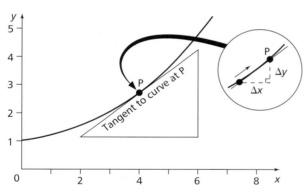

Fig 12.24 Tangents and curves

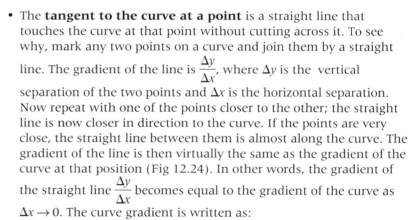

- The **tangent to the curve at a point** is a straight line that touches the curve at that point without cutting across it. To see why, mark any two points on a curve and join them by a straight line. The gradient of the line is $\frac{\Delta y}{\Delta x}$, where Δy is the vertical separation of the two points and Δx is the horizontal separation. Now repeat with one of the points closer to the other; the straight line is now closer in direction to the curve. If the points are very close, the straight line between them is almost along the curve. The gradient of the line is then virtually the same as the gradient of the curve at that position (Fig 12.24). In other words, the gradient of the straight line $\frac{\Delta y}{\Delta x}$ becomes equal to the gradient of the curve as $\Delta x \rightarrow 0$. The curve gradient is written as:

$$\frac{dy}{dx}$$

where $\frac{d}{dx}$ means 'rate of change'.

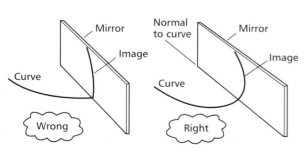

Fig 12.25 Drawing the normal to a curve

- The **gradient of the tangent** is a straight line and is obtained as explained above. Drawing the tangent to a curve requires practice. This skill is often needed in practical work. The **normal** at the point where the tangent touches the curve is the straight line perpendicular to the tangent at that point. An accurate technique for drawing the normal to a curve using a plane mirror is shown in Figure 12.25. At the point where the normal intersects the curve, the curve and its mirror image should join smoothly without an abrupt change of gradient where they join. After positioning the mirror surface correctly, the normal can then be drawn and then used to draw the tangent to the curve.

Turning points

A **turning point on a curve** is where the gradient of the curve is zero. This happens where a curve reaches a **peak** with a fall either side (i.e. a **maximum**) or where it reaches a **trough** with a rise either side (i.e. a **minimum**). Where the gradient represents a physical quantity, a turning point is where that physical quantity is zero. Figure 12.26 shows an example of a curve with a turning point. This is a graph of the vertical height against time for a projectile that reaches a maximum height, then descends as it travels horizontally. The gradient represents the vertical component of velocity. At maximum height, the gradient of the curve is zero so the vertical component of velocity is zero at that point.

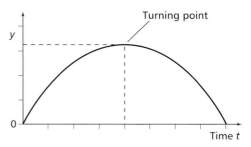

Fig 12.26 *Turning points*

12.6.2 Areas under graphs

The **area** under a line on a graph can give useful information. For example, consider Figure 12.27a which is a graph of the tension in a spring against its extension. Since 'tension × extension' is 'force × distance', which equals work done, then the area under the line represents the work done to stretch the spring.

Figure 12.27b shows a tension v. extension graph for a rubber band. Unlike Figure 12.27a, the area under the curve is not a triangle; but it still represents work done, in this case the work done to stretch the rubber band. (See section 5.5.)

Note The **product** of the *y*-variable and the *x*-variable must represent a physical variable with a physical meaning if the area is to be of use.

Even where the area does represent a physical variable, it may not have any physical meaning. For example, for a graph of p.d. against current, the product of p.d. and current represents power – but this physical quantity has no meaning in this situation.

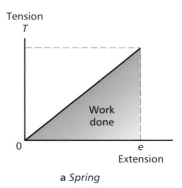

a *Spring*

Examples

More examples of curves where the area is useful include:

- Velocity against time, where the area under the line and the time axis represents **distance moved**.
- Acceleration against time, where the area under the line and the time axis represents **change of velocity**.
- Power against time, where the area between the curve and the time axis represents **energy**.
- Charge and potential difference, where the area between the curve and the p.d. axis represents **energy**.

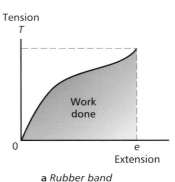

a *Rubber band*

Fig 12.27 *Tension v. extension*

QUESTIONS

1 a Sketch a velocity against time graph (with time on the *x*-scale) to represent the equation

   $v = u + at$, where *v* is the velocity at time *t*.

b What feature of the graph represents:
   (i) the acceleration,   (ii) the displacement?

2 a Sketch a graph of current (on the *y*-axis) against p.d. (on the *x*-axis) to show how the current through an ohmic conductor varies with p.d.

b How can the resistance of the conductor be determined from the graph?

3 An electric motor is supplied with energy at a constant rate:

a Sketch a graph to show how the energy supplied to the motor increases with time.

b Explain how the power supplied to the motor can be determined from the graph.

4 A steel ball bearing was released in a tube of oil and it fell to the bottom of the tube.

a Sketch graphs to show how the following changed with time, for the ball bearing, from the instant of release to the point of impact at the bottom of the tube:
   (i) the velocity,
   (ii) the acceleration.

b (i) What is represented on graph **a**(i) by **1** the gradient, **2** the area under the line?
   (ii) What is represented on graph a(ii) by the area under the line?

Coursework and practical examination

13 Practical skills

13.1 Assessment of experimental skills

13.1.1 Assessment overview

Written tests are also used to assess **Module 1** (Forces and motion) and Module 2 (Electrons and photons). Each of these tests is 30% of the total AS mark. Experimental and investigative skills are assessed in **Module 3**, which includes:

- a written test on waves (which is 20% of the total AS mark),
- practical assessment (which is also 20% of the total AS mark). The method of practical assessment is either through coursework (C2) or a practical examination (C3).

Experimental and investigative skills to be assessed are:

Skill P Planning
- Identify and define the nature of a problem using scientific knowledge and available information.
- Choose effective and safe procedures, select appropriate apparatus and materials, and decide on measurements and observations that are likely to produce reliable and useful results.

Skill I Implementing
- Use apparatus and materials appropriately and safely.
- Carry out practical work safely in a methodical and organised way.
- Make and record detailed observations suitably and make measurements with appropriate precision, using ICT where appropriate.

Skill A Analysing evidence and drawing conclusions
- Communicate scientific evidence and ideas appropriately, including tables, graphs, written reports and diagrams, using ICT where appropriate.
- Recognise and comment on patterns and trends in data.
- Understand what is meant by statistical significance.
- Draw valid conclusions by applying scientific knowledge and understanding.

Skill E Evaluating evidence and procedures
- Assess the reliability and precision of experimental data and the conclusions drawn from it.
- Evaluate experimental techniques and recognise their limitations.

Coursework and practical examination
These skills are all tested either through coursework or through a practical examination. Coursework assessment may be through separate practical

exercises used to test each skill, or via whole investigations. Regardless of whether the skills are assessed through coursework or through the practical exam, such skills need to be developed during the AS course. In addition, practical work is an essential part of acquiring knowledge and understanding of physics. The notes in this chapter are intended to help you to develop your practical skills, not just for assessment purposes, but also to help you to develop your knowledge and understanding of the subject. The subsequent sections in this chapter draw out the key features of each of the above skill areas in such a way as to be of value to every student, whether you are to be assessed through the practical examination or through coursework.

13.1.2 The practical examination (1 hour 30 minutes)

This consists of a planning exercise and a practical test.

The planning exercise

The planning exercise is worth about a quarter of the total mark for the practical examination. The exercise is to be carried out by the candidates over about 7 to 10 days, in the period of two months before the practical test. Sufficient work must be carried out under direct supervision, to allow the teacher to 'authenticate' (i.e. confirm) the work of each candidate as that of the candidate. The planning exercise is collected from candidates on or before the day of the practical test. The planning exercise and the practical test are marked by an external examiner. The mark scheme for the planning exercise matches the mark criteria for the coursework assessment of Skill P (Planning).

The planning exercise is intended to be completed within 7 to 10 days of being issued to candidates. This time can be used to research the topic, using library or other resources, and to carry out any preliminary work. Notes for guidance, as summarised below, are issued with the planning exercise. Many centres require the plan to be written under supervision in a particular lesson.

Summary of notes for guidance

- The plan should have a clear structure. Diagrams should be labelled and positioned appropriately.
- Technical and scientific terms should be used correctly and written in clear and correct English.
- The plan should be about 500 words on A4 paper. It can be hand-written or word-processed.
- The plan should be based on the use of standard equipment, apparatus and other materials available in a school or college science laboratory.
- Sources used in researching the plan should be listed at the end of the plan. References to the sources should be included in the plan, where appropriate.

The practical test

The practical test is designed to test Skills I (Implementing), A (Analysing) and E (Evaluating). The test consists of two compulsory practical questions. One question is longer than the other question and is set in the same general context as the planning exercise, but will not be the same task. The questions are structured in several parts. Instructions are given in each part; candidates are expected to carry out each part and write an answer in the space provided in the question paper. One of the questions may include making a set of measurements and using the measurements to plot a graph. The question might then require some feature of the graph, such as its gradient, to be linked to a given equation. Part of one of the questions will require a lengthy written evaluation. Thus the practical test is a test not just of practical skills, but also of reading and writing skills – requiring rapid and accurate reading as well as the ability to write coherent and fluent prose.

13.1.3 Coursework

The mark 'descriptors' (i.e. criteria) used to judge the mark for your coursework are outlined below. To be awarded the mark at a certain level, all the criteria up to and including that level must be met. An intermediate mark is awarded if only one of the criteria at a certain level is met, provided all the lower criteria are met. If just one lower criteria is not met, your mark plummets to the level of the corresponding lower level.

Skill P: Planning (8 marks maximum)

1 mark

- Define a problem, make a prediction where relevant and plan a fair test or a practical procedure.
- Choose appropriate equipment.

3 marks

- Define a problem, using scientific knowledge or understanding, and identify the key factors to vary or control or take into account.
- Decide on a suitable number and range of observations or measurements to be made.

5 marks

- Make an appropriate plan, using detailed scientific knowledge and understanding and information from preliminary work or a secondary source, taking account of the need for safe working; justify any prediction and produce a clear account using appropriate specialist terms.
- Describe the plan, including choice of equipment and how to produce precise and reliable evidence.

7 marks

- Evaluate and use information from a variety of sources to develop a plan with logical steps, linked to underlying scientific knowledge and understanding; use spelling, punctuation and grammar accurately.
- Justify the plan, including choice of equipment, in terms of the need for precision and reliability.

8 marks

- The plan is exceptional in terms of originality, depth, flair and the use of novel or innovative methods.

Skill I: Implementing (7 marks maximum)

1 mark

- Demonstrate competence and safety awareness in simple techniques.
- Make and record observations and measurements which are adequate for the activity.

3 marks

- Demonstrate competence in practical techniques and manipulate materials and equipment with precision.
- Make and record, clearly and accurately, systematic and accurate observations and/or measurements.

5 marks

- Demonstrate competence and confidence in practical techniques, and adopt safe working practices throughout.
- Make observations and/or measurements with precision and skill, and record them in an appropriate format. Recognise systematic and random errors that affect the reliability and accuracy of the results.

7 marks

- Demonstrate skilful and proficient use of all techniques and equipment.
- Make and record all observations and/or measurements in appropriate detail with the maximum precision possible; respond to serious systematic and random errors by modifying procedures where appropriate.

Skill A: Analysing evidence and drawing conclusions (8 marks maximum)

1 mark

- Process some experimental evidence.
- Identify trends or patterns in the evidence and draw simple conclusions.

3 marks

- Process and present experimental evidence, including graphs and calculations as appropriate.
- Link conclusions drawn from processed evidence with associated scientific knowledge and understanding.

5 marks

- Process and analyse evidence in detail, including, where appropriate: the use of statistics; the calculation of gradients; the plotting of intercepts of graphs.
- Draw conclusions, linked to scientific knowledge, consistent with the processed evidence; produce a clear account with appropriate use of specialist vocabulary.

7 marks

- Make deductions from the processed evidence, using detailed scientific knowledge and understanding, with due regard to terminology and use of significant figures.
- Draw conclusions that are appropriate, well-structured, concise, comprehensive, accurate, and linked coherently to underlying scientific knowledge and understanding.

8 marks

- The work is exceptional in terms of originality, depth, flair and the use of novel or innovative methods.

Skill E: Evaluating evidence and procedures (7 marks maximum)

1 mark

- Make relevant comments on the suitability of experimental procedures.
- Recognise any anomalous results.

3 marks

- Recognise how limitations in the experimental procedures, or plan, may result in sources of error.
- Comment on the accuracy of the observations and/or measurements, suggesting reasons for any anomalous results.

5 marks

- Indicate significant limitations of experimental procedures and/or plan and suggest improvements.
- Comment on the reliability of the evidence and evaluate the main sources of error.

7 marks

- Justify proposed improvements to the procedures and/or plan that would improve the reliability of the evidence and minimise significant sources of error.
- Assess the significance of the uncertainties in the evidence in terms of their effect on the validity of the final conclusions drawn.

When you carry out a scientific investigation, you are entering unknown territory. The investigation might have been carried out many times before, but it's the first time that you have done it – you might notice something that no one else has ever noticed before! Previous investigators might have brushed off such an observation as trivial or irrelevant. You need to plan, and be prepared to change your plan if you discover something unusual that you feel should be followed up. So the first rule of planning a scientific investigation is to make sure your plan doesn't become something you follow with your mind closed down. An inflexible plan for an investigation is little more than a set of instructions, and won't provide much scope to analyse and evaluate evidence. The investigations you carry out should be based on knowledge and understanding from part of the AS specification.

What do I investigate?

Once you have chosen a topic for investigation, you need to make a plan. Before you do that, you need to decide in more detail what you intend to investigate. For example, you might have chosen to investigate factors that affect the resistance of a metal wire. You might decide that one such factor is the temperature of the wire. The problem can then be defined as 'An investigation into the effect of temperature on the resistance of a metal wire'. You could make a prediction about the effect of changing the temperature on the resistance. Your scientific knowledge should be such that you know that the resistance of a wire also depends on its length. Therefore, to make it a fair test, you would need to keep the length of the wire constant.

How do I measure it?

The next step is to think about how you could measure the resistance of the metal wire at different temperatures. So you need to think of a suitable method to change the temperature of the wire, and a method to measure its resistance. You could research different methods of measuring resistance, find out what apparatus is available, and choose a method that makes use of apparatus that is available. Is it sensible to keep the current through the wire constant, or is it best to keep the p.d. across the wire constant? You need to be clear how to use the measurements to calculate the resistance. You would also need to devise a method of changing and measuring the temperature of the wire. One possible way of doing this is to place a coil of the wire in a tube of oil, which would then be placed in a water bath. The temperature of the wire can then be changed by warming the water in the water bath. The temperature of the oil would give the temperature of the wire.

Hints

There seems to be so much to consider, so here are some hints to help you get organised:

- **Research** the topic area, by selective reading from your text book or any other suitable books.
- **Narrow** the area of interest down to an investigation with just a few variables involved. Make sure you have the facilities to control and measure the variables. Perhaps a little more reading-up is necessary here. You could make a prediction about the effect of one variable on another at this stage. Check that your teacher thinks the investigation is feasible in the time available.
- **Note** the variables you intend to measure or control. Plan to operate on a 'cause and effect' basis, with one variable being changed causing another variable to alter; the other variables need to be monitored to make sure they do not change. A variable that can't be kept constant could make the variable being measured alter at random. Some preliminary tests might be needed here. Consider safety issues before you do any preliminary tests.
- **List** the equipment necessary for the measurements. Check it is available when you wish to use it. If necessary, obtain materials from home or from a local supplier.
- **Design**, construct and test any special pieces of equipment you need for your measurements.
- **Modify** your plans or designs, as a result of any more preliminary tests thought to be needed.
- **Write** a clear account of your plans, including relevant theory and your reasons for choices and decisions made.

Let's see how this procedure might work in practice, applied to the investigation of 'The motion of an object falling through a fluid'.

Researching

Researching should enable you to discover that the speed of an object falling through a fluid becomes constant. This speed is called the **terminal speed**. At the terminal speed, the weight of the object is equal to the **drag force** of the fluid on the object. The drag force depends on the **viscosity** of the fluid, which is a measure of how easily the fluid flows. It also depends on the shape and size of the object. For a sphere of radius r moving through a fluid at speed v, Stokes' Law gives the drag force $F = 6\pi\eta r v$ where η is the viscosity of the fluid.

Narrowing down

Narrowing the investigation down could be done by considering what factors might affect the terminal speed of an object falling through a fluid. Such factors include the viscosity of the fluid and the weight, the shape and the size of the object. The shape can be kept the same by using spheres. You could investigate how the terminal speed depends on the radius of different spheres of the same material (e.g. steel ball bearings) in the same fluid, possibly water. You could use Stokes' Law to make a prediction as to how you would expect the terminal speed of a sphere to depend on its radius.

Note that, by using the same material, the weight is taken account of – since the weight depends on the radius. Also, by using different spheres, the shape of the object is the same. The viscosity of a fluid depends on temperature, so the temperature of the fluid needs to be constant. To measure the speed, each sphere needs to be timed over a measured distance.

- Should the **distance fallen** in a certain time be measured, or the **time taken** to fall a certain distance? In other words, should the distance fallen be kept constant and the time measured for spheres of different radii; or should the time of fall be kept the same and the distance fallen measured for each sphere? What justification do you have for choosing one or the other method? If you choose a fixed distance, what distance should it be? What justification do you have for this choice? Some preliminary tests might help here.

- How many ball bearings of different radii should be used? What should be the range of the radii chosen? How many times should the speed of each ball be measured? How many measurements of the diameter of each ball need to be made? Again, you need to justify your choices.

- Remember to take account of safety considerations. Don't handle electrical timing equipment with wet hands! Make sure the cylinder is stable and secure.

Listing equipment

List the equipment needed. This would include a stopwatch, or an electronic timer with light gates if the timing is too short; a metre rule; a tall glass cylinder; elastic bands as markers; a micrometer. A magnet might prove useful to 'fish' each ball out of the tube if you use steel ball bearings. You could also use a top-pan balance to measure the mass of each ball, if you wanted to check they all have the same density.

Designing equipment

Design any special apparatus you might need, e.g. if you use light gates, to ensure the ball does not miss the light beam.

Modifying plans

Modifying your plans might be necessary if a sphere seems to change speed as it falls. You could test this by timing the same sphere over different distances fallen, measured from the point of release of the sphere at the surface. You might find that the spheres need to fall a few centimetres to reach terminal speed and you would then need to position the markers appropriately.

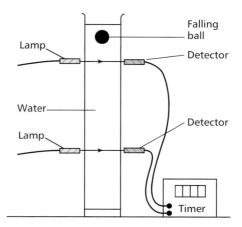

Fig 13.1 *Investigating the motion of a sphere falling in water*

Writing an account

Writing about your plans can be done at different stages during the planning process. If you do this, you need to make sure that it is clear what you propose to do; it should also be clear where any part of the plan has been discarded, with a statement of why that part was not included. However, you could finish up with lots of 'dead-ends' cluttering your plan. If you leave your account of the plan until you are ready to proceed, you might forget some important considerations. You could use a note book to keep track of everything you do, and the decisions you make, as you proceed through the planning stages. Then you could write your account, making use of your notes in the process. Your account should not be more than a few pages of A4, including labelled diagrams. It should include your prediction and the justification for it, as well as how you would use your measurements to test your prediction.

13.3.1   A measurement checklist

At AS level, Skill I: Implementing requires you to manipulate simple apparatus, make accurate and precise measurements and apply recognised safety procedures. You should be able to:

- Measure lengths using a ruler, a vernier scale, a micrometer and callipers.
- Weigh an object, and determine its mass using a spring balance, or a lever balance, or a top-pan balance.
- Use a protractor to measure an angle.
- Measure time intervals using clocks, stopwatches and the time base of an oscilloscope.
- Measure temperature using a thermometer.
- Use ammeters and voltmeters with appropriate scales.
- Use an oscilloscope.

In addition you should be able to:

- Read analogue and digital displays.
- Use calibration curves.
- Distinguish between systematic errors (including zero errors) and random errors.
- Understand what is meant by accuracy, linearity, reliability and precision.

13.3.2   Measurements and errors

Measurements play a key role in science, so they must be reliable. **Reliability** means that a consistent value should be obtained each time the same measurement is repeated. An unreliable weighing machine in a shop would make the customers go elsewhere. In science, you can't go elsewhere, so the measurements must be reliable. Each time a given measurement is repeated, it should give the same value within acceptable limits.

Example

Consider the example of measuring the diameter of a uniform wire using a micrometer. Suppose the following diameter readings are taken for different positions along the wire from one end to the other:

     0.34 mm, 0.33mm, 0.36 mm, 0.33 mm, 0.35 mm

The mean value, $<d>$, is 0.34 mm, calculated by adding the readings together and dividing by the number of readings. If the difference between each reading and $<d>$ changed regularly from one end of the wire to the other, it would be reasonable to conclude that the wire was non-uniform. Such differences are called **systematic** errors. No such differences can be seen in the set of readings. In other words, there is no obvious pattern in the differences so the differences are said to be **random** errors.

What causes random errors?

The instrument might be affected in some way by changes that are not controlled, for example temperature. This is unlikely in the case of a micrometer, as thermal expansion would be insignificant. The experimenter might not use or read the micrometer correctly consistently. This shouldn't happen to a skilled experimenter. In the case of the diameter readings, the wire might have been manufactured with random variations in its diameter along its length. In other words, one or more uncontrolled factors would have affected the machine used to make the wire.

The spread of diameter readings is from 0.33 mm to 0.36 mm. Most of the readings lie within 0.1 mm of the mean value. So the diameter can be written as 0.34 ± 0.1 mm. The diameter is accurate to ±0.1 mm. This estimate of uncertainty is called the **probable error** of the value. Statistics formulas can be used to calculate probable errors.

13.3.3   Using instruments

Instruments used in the physics laboratory range from the very basic (e.g. a millimetre scale) to the highly sophisticated (e.g. a multi-channel data recorder). Whatever type of instrument you use, you need to be aware of the following key points.

Zero error

Does the instrument read zero when it is supposed to? For example, voltmeters and ammeters need to be checked before use to ensure they read zero when disconnected. Another example is a micrometer; when the measuring gap is closed, the reading should be zero. If not, the zero reading must be taken into account when measuring the gap width.

Resolution

The **resolution** of an instrument, sometimes referred to as its precision, is the smallest non-zero reading that can be measured using the instrument. For example, the smallest non-zero reading of a digital voltmeter that has a resolution of 0.001 V is 0.001 V.

- If the reading of an instrument fluctuates when a reading is being made, the measurement cannot be made to the degree of resolution of the instrument. Several measurements would need to be made and the mean value calculated. The **precision** of the measurement is given by the probable error of the readings and can be judged from the spread of the fluctuations.

- If the fluctuations are relatively small, the mean value and probable error can be estimated directly. For example, if the read-out of a digital voltmeter fluctuates between 0.502 and 0.504 V, the measurement can be noted as 0.502 ± 0.002 V. If the reading does not fluctuate, the precision of the measurement is the resolution of the scale.

Linearity

This is a design feature of most analogue instruments. An analogue instrument, such as a moving coil meter, has a pointer which is read against a scale. The scale is linear if its graduations are at equal distances. Provided equal increases of current through the meter cause the pointer to deflect by equal amounts, the reading of the pointer against the scale should be proportional to the quantity being measured.

A lens may be used to read a scale with greater precision, as shown in Figure 13.2. Provided the pointer is thin compared to a scale division, using a lens enables the scale to be read to within 0.2 of a division. In Figure 13.2, the scale can be read to ± 0.02 A ($= 0.2$ of a scale division $\times 0.1$ A).

Measurement errors are caused in analogue instruments if the pointer on the scale is not observed correctly. The observer must be directly in front of the pointer when the reading is made. Figure 13.3 shows how a plane mirror is used for this purpose. The image of the pointer must be directly behind the pointer to ensure the observer views the scale directly in front of the pointer.

Calibration curves

These are used where an instrument is used to measure a physical quantity indirectly. For example, a resistance thermometer makes use of the property that the resistance of a metal wire or a thermistor varies with temperature. The resistance of the wire or thermistor is measured and the corresponding temperature is read off a calibration curve that shows how the resistance varies with temperature. A calibration curve is also used to check the **accuracy** of an instrument. For example, if a resistance thermometer is used to measure a known temperature, the measured resistance should be the same as the resistance read off the calibration curve for that temperature (Fig 13.4).

Accuracy and precision

Accuracy is only possible if there are no systematic errors when a measurement is made. Precise readings are not necessarily accurate readings, because systematic errors could make precise readings lower than they ought to be. For example, suppose the spring of a spring balance becomes weaker due to constant use. Its readings would be too high, because the spring stretches more than it should each time a weight is hung from it. The difference between each reading and what it should be is a systematic error. Precise readings can still be made with the spring balance, but the readings are not as accurate as they should be because there is a systematic error present.

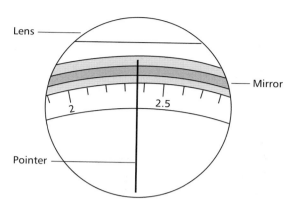

Fig 13.2 *Magnifying a scale*

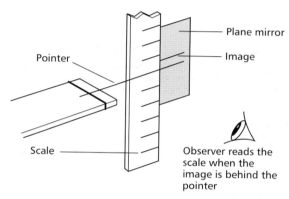

Fig 13.3 *Reading a scale*

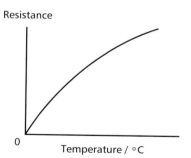

Fig 13.4 *A calibration curve*

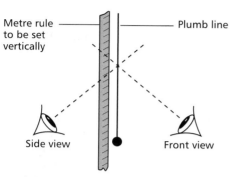

Metre rule to be set vertically

Plumb line

Side view

Front view

If the metre rule appears parallel to the plumb line from the front and the side, the rule must be vertical

Fig 13.5 *Finding the vertical*

13.4.1 Rulers and scales

Metre rulers are often used as vertical or horizontal scales in mechanics experiments. To set a metre ruler in a vertical position, use a **plumb line** (i.e. a small weight on a string) to judge if the rule is vertical. You need to observe the rule next to the plumb line from two perpendicular directions. If the rule appears parallel to the plumb line from both directions, then it must be vertical. To ensure a metre rule is horizontal, use a set square to align the metre rule perpendicular to a vertical metre rule.

13.4.2 Micrometers and verniers

Micrometers

Micrometers give readings to within 0.01 mm. The barrel of a micrometer is a screw with a pitch of 0.5 mm.

* The edge of the barrel is marked in 50 equal intervals so each interval corresponds to changing the gap of the micrometer by $\frac{0.5}{50}$ mm = 0.01 mm.

* The stem of the micrometer is marked with a linear scale graduated in 0.5 mm marks.

* The reading of a micrometer is where the linear scale intersects the scale on the barrel.

* Figure 13.6 shows a reading of 4.06 mm. Note that the edge of the barrel is between the 4.0 and 4.5 mm marks on the linear scale. The linear scale intersects the sixth mark after the zero mark on the barrel scale. The reading is therefore 4.00 mm from the linear scale + 0.06 mm from the barrel scale.

To use a micrometer correctly:

* Check its zero reading and note the zero error if there is one.

* Open the gap, by turning the barrel, then close the gap on the object to be measured. Turn the knob until it slips. Don't overtighten the barrel. (See Fig 13.6.)

* Take the reading and note the measurement after allowing, if necessary, for the zero error.

Note that the precision of the measurement is ±0.010 mm because the precision of the reading and the zero reading are both ±0.005 mm. So the difference between the two readings (i.e. the measurement) has a precision of 0.010 mm.

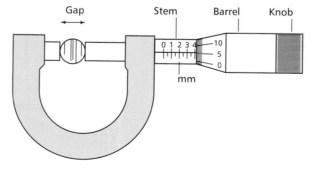

Gap   Stem   Barrel   Knob

mm

Fig 13.6 *Using a micrometer*

Vernier callipers

Vernier callipers are used for measurements of distances up to 100 mm or more. Readings can be made to within 0.1 mm. The sliding scale of any vernier has ten equal intervals, covering a distance of exactly 9 mm, so each interval of this scale is 0.1 mm less than a 1 mm interval. To make a reading:

- The zero mark on the sliding scale is used to read the main scale to the nearest millimetre. This reading is rounded down to the nearest millimetre.

- The mark on the sliding scale closest to a mark on the millimetre scale is located and its number noted. Multiplying this number by 0.1 mm gives the distance to be added on to the rounded-down reading.

Figure 13.7 shows the idea. The zero mark on the sliding scale is between 39 and 40 mm on the mm scale. So the rounded-down reading is 39 mm. The fifth mark after the zero on the sliding scale is nearest to a mark on the millimetre scale. So the extra distance to be added on to 39 mm is 0.5 mm (= 5 × 0.1 mm). Therefore, the reading is 39.5 mm.

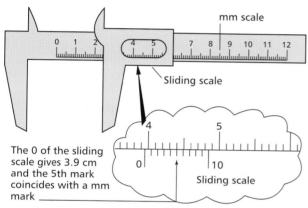

The 0 of the sliding scale gives 3.9 cm and the 5th mark coincides with a mm mark

Fig 13.7 *Using a vernier*

13.4.3 Timers

Stopwatches

Stopwatches used for interval timings are subject to human error because reaction time, about 0.2 s, is variable for any individual. With practice, the delays when starting and stopping a stopwatch can be reduced. Even so, the precision of a single timing is unlikely to be better than 0.1 s. Digital stopwatches usually have read-out displays with a resolution of 0.01 s, but human variability makes such precision unrealistic; the precision of a single timing of a digital stopwatch is the same as an analogue stopwatch.

Timing oscillations requires timing for as many cycles as possible. The timing should be repeated several times to give an average (i.e. mean) value. Any timing that is significantly different to the other values is probably due to miscounting the number of oscillations, so that timing should be rejected. For accurate timings, a **fiducial mark** is essential. The mark acts as a reference to count the number of cycles as the object swings past it each cycle.

Electronic timers

Electronic timers use gates to start and stop an electronic counter. The counter is supplied with 1.0 kHz pulses when the 'condition' of the gates allows. So the counter read-out has a resolution of 1 ms. However, just as with a digital stopwatch, a timing should be repeated, if possible, several times to give an average value.

Electronic timers are often used with **light gates**. A light gate consists of a light source, that produces a light beam, and a light detector connected to the electronic timer. For example:

- **With one light gate**, the speed of a dynamics trolley can be measured by attaching a card to the trolley and making the card interrupt the light beam. The timer measures the time taken by the card to pass through the light beam. The speed is calculated by dividing the length of the card by the time taken for the card to pass through the light beam.

- **With two light gates**, the average speed of an object over the distance between the two light gates can be measured. The electronic timer is used to measure the time taken for the object to move from one light gate to the other. The average speed is the distance between the light gates divided by the time taken by the object to move from one gate to the other.

Light gates may be connected via an interface unit to a microcomputer. Interrupt signals from the light gates are timed by the microcomputer's internal clock. A software program is used to provide a set of instructions to the microcomputer.

13.4.4 Balances

A balance is used to measure the weight of an object. Spring balances are usually less precise than lever balances. Both types of balance are usually much less precise than an electronic top-pan balance. The scale or read-out of a balance may be calibrated for convenience in kilograms or grams. As explained in section 3.1, the weight of an object in newtons is equal to its mass in kilograms × g, where g is the acceleration of free fall. Provided the balance is not moved to a location where g is different, its reading can be converted from newtons to kilograms or grams, or vice versa, using the conversion factor of 9.81 N per kilogram. As a general rule, the reading of an electronic top-pan balance should therefore be tested using **standard masses** (i.e. accurately known masses).

See section 6.3 for the use of ammeters and voltmeters, and section 10.2 for the use of the oscilloscope.

13.5 Data analysis

13.5.1 Data processing

- **For a single reading**, the resolution of the measuring instrument determines the precision of the reading. For example, consider the following measurements of the diameter of a wire:

    0.34 mm, 0.33mm, 0.36 mm, 0.33 mm, 0.35 mm

    The micrometer used to make the measurements has a resolution of 0.01 mm, so each reading has a precision of 0.01 mm.

- **For the mean value of several readings**, the number of significant figures of the mean value should be the same as the precision of each reading. For example, the mean value of the diameter readings above works out at 0.342 mm, but the third significant figure cannot be justified as the precision of each reading is only 0.01 mm. Therefore the mean value is rounded down to 0.34 mm.

As a general rule, when you use a set of measurements in a calculation, don't make the mistake of giving the answer to the calculation to as many significant figures as shown on the calculator display. The same rule applies to a calculation using experimental data. Always give your answer to the **same number of significant figures as the data** you are using. Suppose different measurements in a set of experimental data are measured to differing degrees of precision: your answer should be given to the same number of significant figures as the measurement with the **least** number of significant figures.

Using error estimates

How confident can you be in your measurements, and any results or conclusions you draw from your measurements? No one else will have much confidence in your results if you are very confident in your results because you did the measurements carefully. Such a vague statement will not convince anyone. What you need to do is to estimate and record the probable error of each measurement. If you work out what each probable error is as a percentage of the measurement, the **percentage probable error**, you can then see which measurement is least accurate. You could then go back to the apparatus to see if that measurement can be made more accurately.

Worked example

The mass and diameter of a ball bearing were measured and the probable error of each measurement was estimated.

The mass, m, of a ball bearing $= 4.85 \times 10^{-3} \pm 0.02 \times 10^{-3}$ kg

The diameter, d, of the ball bearing $= 1.05 \times 10^{-2} \pm 0.01 \times 10^{-2}$ m

Calculate and compare the percentage probable error of these two measurements.

Solution

The percentage probable error of the mass $m = \dfrac{0.02}{4.85} \times 100\%$
$= 0.4\,\%$

The percentage probable error of the diameter

$d = \dfrac{0.01}{1.05} \times 100\% = 1.0\,\%$

The diameter measurement is therefore twice as inaccurate as the mass measurement.

More about errors

> **When two measurements are added or subtracted, the probable error of the result is the sum of the probable errors of the measurements.**

Example

If the mass of a beaker is measured when it is empty then when it contains water, and:

- the mass of an empty beaker $= 65.1 \pm 0.1$ g,
- the mass of the beaker and water $= 125.6 \pm 0.1$ g,

then the mass of the water could be as much as:
$$(125.6 + 0.1) - (65.1 - 0.1) \text{ g} = 60.7 \text{ g}$$
or as little as: $\quad (125.6 - 0.1) - (65.1 + 0.1) \text{ g} = 60.3 \text{ g}$

The mass of water is therefore 60.5 ± 0.2 g.

> **When a measurement in a calculation is raised to a higher power, the percentage probable error is increased by the same factor.**

Example

For example, suppose you need to calculate the area A of cross-section of a wire that has a diameter of 0.34 ± 0.01 mm.

You will need to use the formula $A = \dfrac{\pi d^2}{4}$. The calculation should give an answer of 9.08×10^{-8} m². But what is the probable error of A?

The probable error of d is $\dfrac{0.01}{0.34} \times 100\,\% = 2.9\,\%$.

To work out the probable error of A, you could:

- Calculate the area of cross-section for d as $0.34 - 0.01$ mm $= 0.33$ mm. This should give an answer of 8.55×10^{-8} m².
- Calculate the area of cross-section for d as $0.34 + 0.01$ mm $= 0.35$ mm. This should give an answer of 9.62×10^{-8} m².

Therefore, the area lies between 8.55×10^{-8} m² and 9.62×10^{-8} m². In other words, the area is $(9.08 \pm 0.53) \times 10^{-8}$ m² (as $9.08 - 0.53 = 8.55$ and $9.08 + 0.53 = 9.62$).

The percentage probable error of A is $\dfrac{0.53}{9.08} \times 100\% = 5.8\%$.

This is twice the percentage probable error of d. The reason is that the percentage probable error of d^2 is twice the percentage probable error of d.

It can be shown, as a general rule, that for a measurement x:

Percentage probable error in x^n is n times percentage probable error in x

The consequence of this rule is that in any calculation where a quantity is raised to a higher power, the probable error of that quantity becomes much more significant.

13.5.2 Graphs and errors

Straight-line graphs are important because they are used to establish the relationship between two physical quantities. For example, as explained in section 12.5, consider a set of measurements of the distance fallen by an object released from rest and the time it takes. A graph of distance fallen against (time)^2 should be a straight line through the origin. If the line is straight, the theoretical equation $s = \frac{1}{2}gt^2$ (where s is the distance fallen, and t is the time taken) is confirmed. The value of g can be calculated, as the gradient of the graph is equal to $\frac{1}{2}g$. If the straight line does not pass through the origin, there is a systematic error in the distance measurement. Even so, the gradient is still $\frac{1}{2}g$.

A best-fit test

Suppose you have obtained your own measurements for an experiment and you use them to plot a graph that is predicted to be a straight line. The plotted points are unlikely to be exactly straight in line with each other. The next stage is to draw a straight 'line of best fit' so that the points are, on average, as close as possible to the line. Some problems may occur at this stage:

Problem 1

There might be a point much further from the line of best fit than the others. The point is referred to as an **anomaly**. Possible methods for dealing with an anomalous point are as follows:

- If possible, the measurements for that point should be repeated and used to replace the anomalous point if the repeated measurement is much nearer the line.

- If the repeated measurement confirms the anomalous result, there could be a fault in the equipment or the way it is used. For example, in an electrical experiment, an anomaly of the smallest or largest measurement could be caused by a change of the range of a meter to make that measurement.

- If a repeat measurement is not possible, the anomalous point should be ignored and a comment made in your report about it.

Problem 2

The points might seem to curve away from the line of best fit. The probable error of each measurement can be used to give a small range, or **error bar**, for each measurement. Figure 13.8 shows the idea. The straight line of best fit should pass through all the error bars. If it doesn't, the following notes might be helpful. You could use the error bars to draw a straight line of maximum gradient and a straight line of minimum gradient.

- The points lie along a straight line over most of the range, but curve away more and more further along the line. This would indicate that a straight-line relationship between the plotted quantities is valid only over the range of measurements which produced the straight part of the line.

- Only two or three points might seem to lie on a straight line. In this case, it cannot be concluded that there is a linear relationship between the plotted quantities. Further graphs might need to be plotted to find out if a different type of graph would give a straight-line relationship. A data analysis software package on a computer could be used to test different possible relationships.

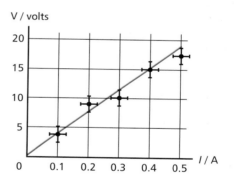

Fig 13.8 Error bars

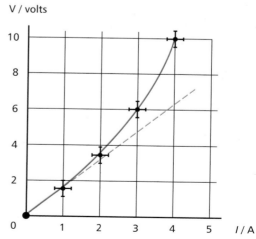

Fig 13.9 Curves

13.6 Report writing

13.6.1 Reports on experiments and short investigations

Before you carry out your first coursework investigation at AS level, you ought to have done sufficient practical work to master the techniques of using laboratory equipment and making accurate measurements. You should also have learned how to analyse data, including plotting and using graphs. Your practical work will probably be directly related to the topics you are studying, giving you a better understanding of these topics. The practical activities you carry out will probably include experiments and short investigations that you have to write a report on. Such reports are important for revision, as knowledge of an experiment or investigation might be tested in one of the module tests. However, they are even more important for another reason, namely helping you to develop the written skills necessary for producing coursework reports.

The experiments and short investigations that precede your coursework investigation might not have required much planning if you were given a set of instructions to follow. You might have been given a practical worksheet telling you how to analyse the data. Even if you were guided 'by hand' through the experiment, writing a report on the experiment is an essential part of practical work for the reasons given above. So here are some guidelines to assist you, when you are writing a report on an experiment or short investigation which has required little or no planning. Coursework reports are considered in the next section (13.6.2).

- Write the descriptive parts of your report in plain language.
- Use capital letters, commas and full stops correctly. Don't make spelling mistakes.
- Write your report using the headings below. This will make your report easy to follow.

Title
Make sure your report has a proper title and a date.

Outline
State the aim of the experiment, or investigation, and give an outline of what is to be measured and how the measurements are to be used to achieve the aim of the experiment.

Apparatus list
Write a list of the apparatus you will use, including the range of any meters listed.

Diagram
Draw a labelled diagram of the apparatus, including its arrangement. Use a pencil and a ruler for this purpose. Make sure you use standard circuit symbols for electric circuits. If you draw more than one diagram, number each diagram (e.g. Figure 1, Figure 2).

Method
Describe how you used the apparatus to obtain your measurements. Your description should include what you did in the order you did it. Include the methods used to control variable quantities and the steps you took to improve accuracy. Be specific and avoid vague phrases. However, don't include trivial details, such as the order in which the components of a circuit were connected together. Where appropriate, you should refer to your diagrams and to the recording of measurements, including any tables of measurements to be given.

Results
Write your measurements in the order you took them. Tabulate your measurements and number each table. If your measurements take up more than a page or so, put them in an appendix to your report and refer to the tables in the results' section. List units and probable errors.

Treatment of results
- Data processing from the measurements, as determined by related theory are to be given here. For calculations on a set of measurements, give the results in a table.
- If there is a graph (or graphs) to be plotted, it should be listed here with a brief account of the reason for plotting this graph and any related theory. The calculation of the gradient of a line can be given on the graph sheet, but the result should be stated here. Comments about the graph should be given.
- Final calculations should be given here, including any necessary theory not described earlier that lies behind the calculations. If the final calculations use data from different parts of the report, the origin of such data in the report should be stated.

Accuracy and errors
Estimates of probable errors should be stated in the measurements section. Identify the most significant sources of error in your measurements, including any systematic errors, and suggest ways in which they could be reduced. Describe the steps taken to deal with anomalous results. Don't repeat the description in the method of steps taken to eliminate random errors.

Conclusions
State the result of your final calculations. Use the degree of accuracy of your measurements as a guide to the number of significant figures in the final result. State any mathematical links between quantities that you have established or verified. Discuss how well you achieved your initial aim; be critical here and don't make a claim without justification.

13.6.2 Coursework reports

If you have developed your report writing skills as you carry out practical experiments and short investigations, you should not have too much difficulty writing a coursework report on a longer practical investigation. There are some important differences though.

Planning

A clear written account of your plans is required. Your teacher will expect you to provide this before you implement your plans. You might have worked out what you intend to do in your head, or on paper, but you must write a clear account of your plans which your teacher will want to read through before you proceed further. Guidance on planning and writing an account of your plans is covered in section 13.2. After your teacher has approved your plan and marked it against the mark criteria for planning (see section 13.1), you should note any suggestions and modify your plan if necessary. You can then include your modified plan as the first part of your coursework report.

Implementing

The next stage is to implement your plan. A further account of your method is not needed, if you implement your plan without having to make any modifications. Your measurements should be listed, or tabulated, in a section entitled 'Results'. Estimates of probable errors should be given here. Take care to ensure all units are correctly given and the measurements are written to an appropriate number of significant figures. See advice on writing up your results in section 13.2. If you need to make modifications to your plans, write some notes at this stage about these modifications, including why it was necessary to make such changes.

Analysing evidence

Under the heading 'Treatment of results', use your measurements for data processing, graph plotting, related graph work and final calculations. This is where practice via previous experiments is likely to prove very valuable. See the section on 'Treatment of results' in 13.6.1. The final result of your calculations should be given here.

Evaluating evidence and procedures

Under the heading 'Evaluation', this is where you can write about accuracy and errors. In addition, you need to discuss the reliability of the data and suggest improvements, where appropriate, that could be made to improve the reliability. To evaluate the main sources of error, you could work out percentage probable errors in your measurements here and then identify the most significant sources of error in your measurements. You should include a discussion of how to reduce the most significant sources of error. Also, discuss any systematic errors that you have identified and suggest ways in which they could be reduced. Steps taken to deal with anomalous results could be described here and steps taken to eliminate or reduce random errors, if such a discussion has not been given earlier.

Drawing conclusions

Under the heading 'Conclusions', restate the result of your final calculations. As explained in section 13.6.1, use the degree of accuracy of your measurements as a guide to the number of significant figures in the final result. State any mathematical links between quantities that you have established or verified. The validity of your conclusions should be discussed in terms of the **strength** of the experimental evidence (e.g. measurement errors, anomalies, 'best fit' difficulties) that you used to draw your conclusions. Proposed improvements to the stategy or experimental procedures could be given here, referring to the above discussion as justification for the proposals.

When you have completed your coursework report, check that you have met as many of the coursework criteria as possible. Make sure there are no gaps at the lower mark levels. In addition, check units, significant figures, spelling, punctuation and grammar before you hand your report in to your teacher.

With your GCSE course successfully completed, don't sit back and think that your AS course is going to be relatively easy because you are taking just your best subjects now. You may well have already discovered that you need to put in a considerable amount of extra time outside lessons to cope with the demands of your AS course. If you are studying four AS subjects and other courses, such as General Studies or Key Skills, you have a very busy programme and can't afford to slip behind or drift. The standard of AS level examinations is based on the level of understanding and knowledge developed after one year of further study beyond GCSE. In fact, the AS examinations are taken after about 32 weeks of study. So you need to:

- Be organised and stay organised.
- Keep your notes in good order, and make sure they are clear and readable.
- Make sure you understand each topic and learn it as you go along.
- Keep up with homeworks and coursework.
- Revise each topic thoroughly when you have class tests.
- Use feedback from homeworks and class tests to pinpoint and overcome gaps in your knowledge.
- Revise thoroughly for module tests well in advance.

Here are some hints and tips to help you develop your study skills in Physics. No doubt some of the points will also apply to your other subjects.

Note taking

Note taking in class is for your benefit; it should form the basis of your learning in the subject and the basis of your revision. Your teacher might write notes on the board for you to copy, or provide you with printed notes or notes in an electronic file. Don't forget to date your notes and use clear headings, so you can see later on where each topic starts and finishes. If such notes are not clear, use this text book to read about the topic while it is fresh in your mind; then write a clear set of notes covering all the points in the notes you took in class. Likewise, if you have to make your own notes, you will find this text book invaluable. However, don't just rewrite everything you read, as you need to keep your notes brief. Highlight key words, use diagrams and colour where possible, make lists of the main points and equations, and link them together on topic charts.

Learning

Learning in class about a new topic needs to be followed up by private study. If you don't understand something in class, ask the teacher either in class or afterwards. Homeworks set by teachers are to check that you understand a topic and to provide feedback to you. Before you tackle a piece of homework, read through your notes and the relevant part of the text book. If you are unable to complete a piece of homework within a reasonable time, put it aside and ask for help from your teacher. By tackling a piece of homework well in advance of the deadline, you give yourself that bit of extra time to seek help if you need it. Physics homework questions are often short structured calculations that reinforce theory work in class. You gain knowledge about a topic and its applications through such questions. However, to understand a topic thoroughly, you need to be able to:

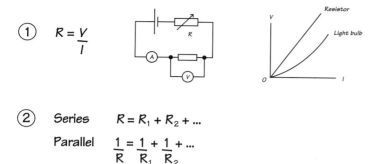

Fig 14.1 *Making notes*

- Explain the key ideas clearly in your own words.
- See how the key ideas relate to experimental and practical work in the topic.
- Fit the key ideas into the theoretical framework of the topic.

The practical work you do in class, the demonstrations you see, the computer exercises you do, and the theory you cover in class are all designed to help you to understand the **key ideas** of a topic. Once you understand the key ideas, you can then see their relevance in terms of **applications**.

To show that you understand the key ideas of a topic, you need to be able to explain the ideas in your own words. Homework questions that require descriptive answers are designed to test your ability to explain key ideas and, therefore, they test your grasp of those ideas. You need to take such questions seriously; don't just dash your answers off. Think carefully about what you intend to write before you do so, and be specific. In other words, answer the question asked and don't waffle!

Read around each topic

Read around each topic using the text book, library books and relevant scientific articles. Use the Internet to find out more about topics that catch your interest. Such reading will help you to develop your writing skills, as well as boosting your knowledge and confidence. In addition, you should find that increased awareness of the relevance of the subject helps to shape up your thoughts to do with careers, at a time when you need to start thinking about what to do after completing your two-year A level course.

Revisiting topics

Revisiting topics as you progress through your course is essential. If you don't go over topics regularly, you will probably find your knowledge of them fades and you can't build on them to develop your understanding of new topics. After you learn a new topic, your recollection of the key ideas, facts and equations will probably need to be refreshed periodically. This is why clear and brief notes are important, as well as lists and charts. By reviewing a topic a

week or so later, then a month or so later, then a term later, you refresh your knowledge and understanding of its key points.

Perhaps you have marvelled at people with photographic memories: able to recall facts, figures, equations, stories, poems, etc. How do they do it? If you are working on a topic every day, you soon get to a position where you can recall everything about it. If you are moving on from one topic to another at a fast pace, as you do at AS level, you need to revisit each topic periodically, making use of shortcuts to help you remember (e.g. Roy G Biv for the colours of the spectrum) and testing yourself by writing down the key points from memory.

Data and formulae

The specification lists the data and formulae to be supplied with each question paper and the formulae that must be recalled. All this information is provided on pp. 182–183. The list of formulae to be recalled in the specification covers all the modules, so it has been provided module by module. Each module list consists of all the formulae in the module specification that are not in the list to be supplied with every examination paper. The formulae to be recalled in these module lists should be committed to memory as each topic is covered.

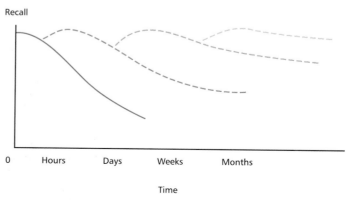

Fig 14.2

Revising and preparing for exams

As you approach tests and examinations, you need to make a revision schedule and stick to it. You will have to revise all your subjects over the same period, as you approach AS module tests, and you may also have coursework or practical work deadlines to meet. So here are some rules for revision:

Time

Make sure you complete your coursework report in good time, before the coursework deadline set by your teacher. This will ensure you don't lose valuable revision time as you get to the end of the course. If you are taking the practical examination, you need to allow time in your schedule for the planning exercise, which may have to be done over the same period as your revision for the module tests.

In-depth cycle

Aim to carry out a first in-depth cycle of revision, covering every topic over a period of a few months. You might well be completing the course and learning new topics during this time, but if you revisited the earlier topics once or twice before this revision cycle then this first cycle of revision should be painless. Use the examination-style questions from the end of each chapter to test your knowledge and understanding of each topic.

Shorter cycle

Work through a second shorter cycle of revision of all your topics, once you have finished the course, to check your knowledge and understanding is secure. Use past papers from your teacher, and use mark schemes to check your answers if they are available. Work through all the questions, including those that require descriptive answers. Physics students do have a tendency to neglect the skills needed to answer questions that require key ideas to be explained in writing.

Final revision

Make a final revision schedule, once you have your exam dates, so you can maximise your efforts for each exam over the period when you are sitting the exams. Aim to go over the key points of each topic during this time, to keep those points fresh in your mind when you go into the exam room. If you have done two previous cycles of revision before, this final part will be no more than confirming all the revision work done previously.

14.2.1 How to revise

How you revise depends on your situation and your temperament:

- Your revision is likely to be more effective if you have a definite aim in mind for each revision session, so you need to choose a topic for each revision session.

- Bear in mind that long revision sessions can be unproductive; you will probably gain more out of two or three short sessions, instead of a long session of the same overall duration.

- Make notes of your notes, list the relevant equations and key words and do small sketch graphs and diagrams. Writing all these features in brief on a single side of A4 provides you with your own revision checklist on that topic. Ask a friend or relative later to ask you questions about your notes to check your recall.

- Pinpoint the key ideas in each topic, and write a brief account of each. You could work with other Physics students to test your understanding by quizzing each other.

- Go over your reports on experiments, to make sure you know the theory of each experiment and how it was carried out.

- Look back over previous tests and homeworks, to make sure you know where you didn't score full marks.

- Work through past papers and other questions that you have not done before. In the first cycle of revision, work through questions by topic as you revise each topic. Then tackle relevant past papers after that, or at the second stage, answering the questions in the appropriate time without using your notes.

14.3 Answering exam questions

All the questions on the papers of Forces and Motion, Electrons and Photons and Wave Properties are compulsory. The first two of these papers are each of 1 hour duration and worth 60 marks. About 50 marks are for structured questions and the rest for the questions that require extended answers. The paper on Wave Properties is of 45 minutes duration and is worth 45 marks. Some questions are structured and some require extended answers. See p. 163 for information about the practical exam.

14.3.1 Structured questions

A structured question has several related parts, which probe your knowledge and understanding of a topic, or part of a topic, in the specification for the module. No questions are set on topics outside the module specification. So, what is in a structured question?

- A structured question often starts with some information, that may be accompanied by a diagram. This is the stem of the question; you must read it carefully, otherwise you might miss the point of the question. It may also contain data to be used later.

- The stem is usually followed by several parts, each consisting of one or more specific questions. Each question is usually short and direct, followed by an answer space to write your answer in. The amount of space is a guide as to how much to write. In other words, if the answer space consists of four lines, then there is no point in writing an essay for that part of the question.

- The number of marks for each part of a structured question is written in the right-hand margin. This is important information, because there is usually one mark awarded for each correct relevant point. In a descriptive answer, the mark scheme might consist of more key points than marks available, so full marks can be scored without writing every key point down.

- Make sure symbols used in equations are defined, where necessary. Standard symbols, such as h for the Planck constant or F for force, do not need defining, but symbols for quantities relating to specific situations do (e.g. d for the diameter of a wire and L for its length). Formulae from the specification lists do not need to be defined symbol by symbol, but should always be quoted at the start of any calculation.

- In a calculation, there are usually the same number of marks as there are key points. Marks in the mark scheme are awarded for the steps in a calculation; correct units must be given, or one mark per question can be lost. Numerical answers given to an inappropriate number of significant figures are penalised (one mark per paper can be lost). Beware writing down a whole number as an answer when it should be followed by a decimal point and several zeroes (e.g. 8 instead of 8.00 is a significant figure error).

- A numerical mistake in one part of a question (e.g. incorrect reading of data) causes the loss of a mark in that part but, provided the rest of the answer is satisfactory, the other marks would still be gained. However, if a final numerical answer is clearly absurd, as a result of an earlier numerical error, the final mark would not be gained.

- An error in the physics of an answer will cause the loss of the marks from that point in that part of the question. For example, using $f = \dfrac{1}{\lambda}$ instead of $f = \dfrac{c}{\lambda}$ is a physics error, so the mark for the frequency calculation would be lost as would the subsequent mark for using $E = hf$ if the use of $E = hf$ is next.

Example of a structured question (with a model answer)

Figure 14.3 shows the dimensions of a wire of diameter 0.22 mm and of length 0.50 m.

The wire is made of material of resistivity $5.2 \times 10^{-8}\ \Omega$ m.

a Calculate the resistance of the pencil lead.

$$\text{Area of cross-section} = \frac{\pi}{4}(2.2 \times 10^{-4})^2 = 3.8 \times 10^{-8}\ \text{m}^2$$

$$\rho = \frac{RA}{l} \text{ so } R = \frac{\rho l}{A} = \frac{5.2 \times 10^{-8} \times 0.5}{3.8 \times 10^{-8}} = 0.68\ \Omega \tag{4}$$

b A student connects the ends of the wire to a battery. The potential difference across the ends is 3.0 V. Calculate the current through the wire.

$$I = \frac{V}{R} = \frac{3.0}{6.8} = 0.44\ \text{A} \tag{2}$$

Diameter

Length

Fig 14.3

14.3.2 Questions that require extended answers

Such a question is usually a separate question that is also used to assess quality of written communication (QWC). The question itself may be relatively short, with little guidance as to specific points. However, it will contain key words or phrases such as 'Explain why...' or 'Describe the... '. Key words such as 'List...' or 'Outline...' are self-explanatory, and will probably not appear where an extended answer is required. Table 14.1 offers some guidance about some of the key words that might appear in questions requiring extended answers.

Writing an extended answer

This calls for considerable thought before putting pen to paper.

Read the question carefully, taking **note** of what you are asked to do as well as the technical phrases used. **Underline** important technical words, to confirm that you have made a mental note of each one, and **think** about the specific demands of the question. For example, if you are asked to compare the behaviour of elastic and plastic materials, you need first to explain what an elastic material is and what a plastic material is. Then you can write about their differences.

The main points to be included in an extended answer could be jotted down in the margin, as a **plan** on the basis of a paragraph on each point. Each point should be described, or explained clearly and concisely, in **specific terms**. Avoid ambiguities, and check there are no contradictions in what you write down; otherwise the examiner will conclude that you are unsure, and no marks will be awarded on the contradictory points. For example, if you write down that 'photoelectric emission occurs if the frequency of a photon exceeds the threshold frequency of the metal' and later on you write that 'the wavelength must be greater than a minimum value', then the second statement being incorrect demonstrates that you do not know the facts, because it contradicts the first statement.

Practise writing extended answers. You can't expect to do well at writing a two-page essay in an exam, if you haven't obtained plenty of practice at such writing. Reading scientific text from this book, or from science magazines, or from broadsheet newspapers, is an important part of this preparation too; you will find you remember phrases, and your subconscious memory will take in the style and formal nature of essay writing, when you read such material.

Remember two marks are awarded on every paper for your **quality** of written communication. This will usually be assessed on your extended answer in each paper. Correct spelling, punctuation and grammar, as well as legible hand-writing, are expected for both marks. Finally, do **not** use text-messaging language. Good luck with your exams!

Table 14/1

Key word	Meaning	Good example	Poor example
Define	State the formal definition.	Speed is distance travelled per unit time.	Speed is how fast an object goes.
Describe	State the main features, with a diagram if appropriate.	The fringe pattern consists of alternate parallel bright and dark fringes that are equally spaced.	The fringe pattern consists of bright and dark fringes.
Explain	Demonstrate your understanding about the theory of a topic.	The drag force is the force of air resistance on the falling object. This force increases as the speed increases. A falling object reaches terminal speed when the drag force is equal and opposite to the weight of the object.	A falling object reaches terminal speed when it can't go any faster.
Discuss	Identify the main points and write about them.	A seat belt is an important safety feature of a vehicle because it prevents the wearer from colliding violently with the vehicle frame in a crash. In a front-end impact where the vehicle suddenly slows down,...	A seat belt stops the wearer from injuring himself or herself by crashing into the vehicle.
Suggest	There may be more than one right answer, so put forward your ideas/views about the issue.	The output power of an electric motor depends on the current and the strength of its magnet. There is a limit to the output power of an electric motor, because too much current through the armature would make it overheat. Also, the magnet of the motor is limited in its strength.	The output power of an electric motor is limited because it stops working if it is overloaded.

Reference section

Data and formulae

Useful data for AS Physics

Acceleration of free fall	$g = 9.81 \text{ m s}^{-2}$
Magnitude of the charge of the electron	$e = 1.6 \times 10^{-19} \text{ C}$
Rest mass of the electron	$m_e = 9.11 \times 10^{-31} \text{ kg}$
Rest mass of the proton	$m_p = 1.67 \times 10^{-27} \text{ kg}$
Speed of light in free space	$c = 3.00 \times 10^8 \text{ m s}^{-1}$
The Planck constant	$h = 6.63 \times 10^{-34} \text{ Js}$

AS Physics formulae supplied in the question paper

Uniformly accelerated motion

$$s = ut + \tfrac{1}{2}at^2$$
$$v^2 = u^2 + 2as$$

Refractive index

$$n = \frac{1}{\sin c} \text{ where } c = \text{critical angle}$$

AS Physics formulae *not* supplied in the question paper

Module 1 Forces and motion

Acceleration

$$a = \frac{(v - u)}{t}$$

Density

$$\rho = \frac{m}{V}$$

Force

$$F = ma$$

Kinetic energy

$$E_K = \tfrac{1}{2}mv^2$$

Moment of a force, or torque of a couple

$$T = Fd$$

Potential energy

$$\Delta E_p = mg\Delta h$$

Power

$$P = \frac{W}{t}$$

Pressure

$$p = \frac{F}{A}$$

Speed

$$v = \frac{s}{t}$$

Strain

$$\text{Strain} = \frac{\Delta l}{l}$$

Stress

$$\text{Stress} = \frac{F}{A}$$

The Young modulus

$$E = \frac{\text{stress}}{\text{strain}} = \frac{Fl}{A\Delta l}$$

Weight

$$W = mg$$

Work done

$$W = Fs$$

Module 2 Electrons and photons

de Broglie equation

$$\lambda = \frac{h}{mv}$$

Electric current

$$I = \frac{\Delta Q}{\Delta t}$$

Electrical energy	$W = IV\Delta t$
Electrical resistance	$R = \dfrac{V}{I}$
Force on a current-carrying conductor	$F = BIl$
Photoelectric effect	$(E_{\text{Kmax}} =)\tfrac{1}{2}mv^2{}_{\text{max}} = hf - \Phi$
Photon energy	$E = hf$
Potential difference	$V = \dfrac{W}{Q},\ V = \dfrac{P}{I}$
Potential divider	$V_1 = \dfrac{V_0 R_1}{(R_1 + R_2)};\ V_2 = \dfrac{V_0 R_2}{(R_1 + R_2)}$
Power	$P = I^2 R = \dfrac{V^2}{R}$
Resistivity	$\rho = \dfrac{RA}{l}$
Resistors in parallel	$\dfrac{1}{R} = \dfrac{1}{R_1} + \dfrac{1}{R_2} + \ldots$
Resistors in series	$R = R_1 + R_2 + \ldots$

Module 3 (Part 1) Wave properties

Wave speed	$v = f\lambda$
Double-slit interference	$\lambda = \dfrac{ax}{D}$

Refraction of light:

1 For light travelling at speed c_0 in air (or a vacuum) passing into a transparent substance, where its speed is c, $n = \dfrac{\sin i}{\sin r}$, where $n = \dfrac{c_0}{c}$

2 For light travelling from a substance of refractive index n_1 into a substance of refractive index n_2, $n_1 \sin i = n_2 \sin r$, where $n_1 = \dfrac{c_0}{c_1}$ and $n_2 = \dfrac{c_0}{c_2}$

Glossary

acceleration of free fall acceleration of an object acted on only by the force of gravity

acceleration change of velocity per unit time

amplitude maximum displacement of a vibrating particle; for a transverse wave, it is the distance from the middle to the peak of the wave

antinode fixed point in a stationary wave pattern where the amplitude is a maximum

base units the five units that define the SI system

braking distance the distance travelled by a vehicle in the time it takes the brakes to stop the vehicle

brittle snaps without stretching or bending when subject to stress

centre of gravity point where the weight of a body may be considered to act

charge carriers charged particles that move through a substance when a p.d. is applied across it

coherence two sources of waves are coherent if they emit waves with a constant phase difference

couple pair of equal and opposite forces acting on a body but not along the same line

cycle interval for a vibrating particle (or a wave) from a certain displacement and velocity to the next time the particle (or the next particle) has the same displacement and velocity

density of a substance mass per unit volume of the substance

diffraction spreading of waves on passing through a gap or near an edge

dispersion splitting of a beam of white light by a glass prism into the colours of a spectrum

displacement distance in a given direction

drag force the force of fluid resistance on an object moving through the fluid

ductile stretches easily without breaking

efficiency the ratio of the work done by a machine to the energy supplied to it

effort the force applied to a machine to make it move

elastic limit point beyond which a wire is permanently stretched

electrolyte a solution or molten compound that conducts electricity

electrolysis process of electrical conduction in a solution or molten compound due to ions moving to the oppositely charged electrode

electromotive force (e.m.f.) the amount of electrical energy per unit charge produced inside a source of electrical energy

electron volt amount of energy equal to 1.6×10^{-19} J, defined as the work done when an electron is moved through a p.d. of 1 V

endoscope optical fibre device used to see inside cavities

equilibrium state of an object when at rest or in uniform motion

error bar representation of a probable error on a graph

error of measurement uncertainty of a measurement

frequency the number of cycles of a wave that pass a point per second

gravitational field strength force of gravity per unit mass on a small object

Hooke's Law the extension of a spring is proportional to the force needed to extend it

interference formation of points of cancellation and reinforcement where two coherent waves pass through each other

internal resistance resistance inside a source of electrical energy

ion a charged atom

kinetic energy the energy of an object due to its motion

Kirchhoff's First Law at a junction, the total current in = the total current out

Kirchhoff's Second Law the sum of the e.m.fs round a complete loop in a circuit = the sum of the p.ds round the loop

light-dependent resistor resistor which is designed to have a resistance that changes with light intensity

linear two quantities are said to have a linear relationship if the change of one quantity is proportional to the change of the other

line of force of a magnetic field line along which a plotting compass points in a magnetic field

load the force to be overcome by a machine when it shifts or raises an object

longitudinal waves waves with a direction of vibration parallel to the direction of travel of the waves

magnetic flux density force per unit current per unit length on a current-carrying conductor placed at right angles to the magnetic field lines

mass measure of the inertia or resistance to change of motion of an object

matter waves the wave-like behaviour of particles of matter

moment of a force about a point force $\times$ perpendicular distance from line of action of force to the point

motive force the force that drives a vehicle

multi-path dispersion the lengthening of a light pulse as it travels along an optical fibre due to rays that repeatedly undergo total internal reflection, having to travel a longer distance than rays that undergo less total internal reflection

Newton's First Law an object remains at rest or in uniform motion unless acted on by a resultant force

Newton's Second Law for constant mass resultant force = mass $\times$ acceleration

node fixed point of no displacement

Ohm's Law the p.d. across a metallic conductor is proportional to the current through it provided the physical conditions do not change

optical fibre a thin flexible transparent fibre used to carry light pulses from one end to the other

path difference the difference in distances from two coherent sources to an interference fringe

period of a wave time for one complete cycle of a wave to pass a point

phase difference the fraction of a cycle between the vibrations of two vibrating particles, measured either in radians or degrees

photoelectricity emission of electrons from a metal surface when the surface is illuminated by light of frequency greater than a minimum value, known as the threshold frequency

photon packet or 'quantum' of electromagnetic waves

plastic deformation deformation of a solid beyond its elastic limit

polarised waves transverse waves that vibrate in one plane only

potential energy the energy of an object due to its position

power rate of transfer of energy

pressure force per unit area acting on a surface perpendicular to the surface

probable error estimate of the uncertainty of a measurement

projectile a projected object in motion acted on only by the force of gravity

radian measure of an angle defined such that 2π radians = $360°$

random error error of measurement with no obvious cause

refraction change of direction of a wave when it crosses a boundary where its speed changes

resistance $\dfrac{\text{p.d.}}{\text{current}}$

resistivity resistance per unit length $\times$ area of cross-section

scalar a physical quantity with magnitude only

semiconductor a substance in which the number of charge carriers increases when its temperature is raised

SI system the scientific system of units

speed change of distance per unit time

stationary waves wave pattern with nodes and antinodes formed when two or more progressive waves of the same frequency and amplitude pass through each other

stopping distance thinking distance + braking distance

strain extension per unit length of a solid when deformed

stress force per unit area of cross-section in a solid perpendicular to the cross-section

superposition the effect of two waves adding together when they meet

systematic error error of measurement with a known cause

terminal speed the maximum speed reached by an object when the drag force on it is equal and opposite to the force causing the motion of the object

thermistor resistor which is designed to have a resistance that changes with temperature

thinking distance the distance travelled by a vehicle in the time it takes the driver to react

threshold frequency minimum frequency of light that can cause photoelectric emission

torque of a couple force × perpendicular distance between the lines of action of the forces

total internal reflection a light ray travelling in a substance is totally internally reflected at a boundary with a substance of lower refractive index, if the angle of incidence is greater than a certain value known as the critical angle

transverse waves waves with a direction of vibration perpendicular to the direction of travel of the waves

vector a physical quantity with magnitude and direction

velocity change of displacement per unit time

wavelength distance between two adjacent wave peaks

weight the force of gravity acting on an object

work force × distance moved in the direction of the force

work function of a metal minimum amount of energy needed by an electron to escape from a metal surface

Answers to numerical questions

Chapter 1

1.1
1 a 63 km   **b** 18°
2 a 40 m s^{-1} East, 69 m s^{-1} North   **b** 21 km
3 6.1 kN vertically up, 2.2 kN horizontal
4 a (i) 10.4 km   (ii) 6.0 km
   b (i) 20.4 km   (ii) 10.0 km
   c 22.7 km
5 a 3.7 N at 33° to 3.1 N
   b 17.1 N at 21° to 16 N
   c 1.4 N at 45° to 3 N and to 1 N
6 a 14.0 N   **b** 6.0 N   **c** 10.8 N

1.2
1 a 80 km h^{-1} **b** 22 m s^{-1}
2 a 2.5 × 10^4 km h^{-1} **b** 1.1 × 10^3 m s^{-1}
3 a 45 000 m   **b** 28.3 m s^{-1}
4 b (i) 4.0 km
     (ii) 30 m s^{-1} then, 25 m s^{-1} in the opposite direction

1.3
1 a 1.5 m s^{-2} **b** −2.5 m s^{-2}
2 a 0.45 m s^{-2} **b** 7.9 m s^{-1}
3 b 0.60 m s^{-2}, 0, −0.40 m s^{-2}

1.4
1 a 2.0 m s^{-2} **b** 220 m
2 a 43 s   **b** −0.93 m s^{-2}
3 a (i) 0.2 m s^{-2} (ii) 90 m
   b (i) −0.75 m s^{-2} (ii) 8.0 s
   d 3.0 m s^{-1}
4 a 5.0 m s^{-2} **b** 7.5 m   **c** 18 m   **d** 6.4 m s^{-1}

1.5
1 a 4.0 m   **b** 8.8 m s^{-1}
2 a 3.2 s   **b** 31 m s^{-1}
3 a (i) 3.9 s   (ii) 38 m s^{-1}
4 a 1.6 m s^{-2} **b** 3.6 m s^{-1} **c** 0.64 m

1.6
1 a (i) 83 s   (ii) 127 s
2 a 600 s
3 b (i) 5750 m   (ii) 10 250 m
4 a (i) 0.61 s   (ii) 5.9 m s^{-1} (iii) 0.43 s   (iv) 4.2 m s^{-1}

1.7
1 a (i) 52 s (ii) 0.49 m s^{-2}
   b (i) 406 m (ii) −1.04 m s^{-2}
2 a 1.5 m   **b** −0.13 m s^{-2}
   c 0.67 m s^{-1} downwards, 13.4 m from the start
3 a (i) 80 m   (ii) 8.0 m s^{-1}
   b (i) 65 s   (ii) −0.12 m s^{-2}
4 a (i) 180 m s^{-1} (ii) 2.7 km
   b 4.4 km
   c 290 m s^{-1}

1.8
1 a 32 m s^{-1} **b** 2.8 s   **c** 39 m
2 a 3.0 s   **b** 49 m
   c 34 m s^{-1} (at 62° to the horizontal)
3 a 0.20 s   **b** 11.7 m s^{-1}
4 a 354 m   **b** (i) 1020 m   (ii) 1020 m   **c** 146 m s^{-1}

1.9
1 a 470 mm   **b** 3.0 m s^{-2}
   c 2.7 m s^{-1} (at 79° to the vertical)

2 a 25.8 m   **b** 2.3 s   **c** 4.6 s   **d** 179 m
3 a 3.5 m s^{-1}, 3.0 m s^{-1} **b** (i) 150 m   (ii) 20 m
4 a 21 m s^{-1}, 10 m s^{-1} **b** 8.7 m   **c** 27 m s^{-1}

Chapter 2

2.1
1 a 7.3 N   **b** 7.3 N at 31.5° to vertical
2 a (i) 2.7 N   (ii) 4.7 N
3 a 139 N   **b** 95 N
4 a 73°   **b** 6.8 N

2.2
1 300 N   **2 b** 6.2 N   **3** 0.27 m   **4** 6.75 N

2.3
1 0.51 N at 100 mm mark, 0.69 N at 800 mm mark
2 a 122 N at 1.0 m end and 108 N from the other end
3 620 kN, 640 kN
4 a 100 N, 50 N   **b** 150 N

2.4
2 89 N   **3 a** 42°

2.5
1 a 50 N   **b** 250 N   **2 a** 1800 N m   **b** 2250 N
3 6 kN   **4** 10.8 kN

2.6
1 a 15 N   **b** 3.0 N   **c** 10.8 N   **2** 7 N
3 a 6.8 N   **b** 52°   **4** 18.0 N
5 a 17 kN   **b** 17 kN
6 a 6.2 N   **b** 11.2 N
7 50 mm away from pivot
8 a 6.8 N   **b** 9.8 N
9 a 2200 N   **b** 3100 N
10 b 950 N at X , 750 N at Y
11 a 2820 kN   **b** 1660 kN and 1540 kN
12 a 8.0 N and 16 N   **b** 38 N and 76 N
13 b 11 kN, 11 kN   **14 b** 28.4 N

Chapter 3

3.1
1 a 0.24 m s^{-2} **b** 190 N   **c** 0.024
2 a 2.4 m s^{-2} **b** 12 000 N
3 a 360 N   **b** 23 s   **4 a** -1.3×10^5 m s^{-2} **b** 260 N

3.2
1 a 5400 N   **b** 7700 N
2 a 60 N   **b** 270 N
3 a 11.8 kN   **b** 11.8 kN   **c** 12.2 kN   **d** 11.3 kN
4 a 1.4 m s^{-2} **b** 12.5 N

3.3
1 a (i) 0.04 m s^{-1} (ii) 1.5 N
3 a 0.14 m s^{-2} **b** 520 m

3.4
1 a (i) 7.2 m   (ii) 33.7 m   **b** 4.1 m
2 a 6.75 m s^{-2} **b** 6750 N   **4 b** 76 m

3.5
1 a 2.0 g **b** 24 kN   **2 a** 80 ms   **b** 375 kN
3 a 0.75 m s^{-2} **b** 675 N   **4 b** 6.4 g

Chapter 4

4.1
1 a 200 J   **b** 4.5 J   **2 a** 48 J   **b** 24 J   **c** 0
3 a 1000 J   **b** 600 J   **c** 400 J
4 a 2.4 N   **b** 0.12 J

4.2
1 a 9.0 J   **b** 9.0 J   **c** 1.8 m
2 a (i) 2.9 J (ii) 2.4 J   **b** 0.5 J
3 a (i) 16 kJ (ii) 5.8 kJ   **b** (i) 10.2 kJ   (ii) 20 N
4 a 590 kJ   **b** 2.4 kJ   **c** 470 kJ   **d** 122 kJ   **e** 1.6 kN

4.3
1 a 11 kJ   **b** 610 J s^{-1} **2** 500 MW
3 a 156 MJ   **b** 140 MJ   **c** 12 MW   **4** 122 m

4.4
1 a 450 J   **b** 1800 J s^{-1}
2 a 480 J   **b** 50 J   **c** 10%
3 a 570 MJ s^{-1} **b** 6.2×10^5 kg
4 a 600 s   **b** 3.7 MJ   **c** 8 %

Chapter 5

5.1
1 a 8.0×10^{-4} m^3 **b** 3.1×10^3 kg m^{-3}
2 a 6.3 kg   **b** 2.0×10^{-3} m^3 **c** 3.1×10^3 kg m^{-3}
3 a 9.6×10^{-6} m^3 **b** 7.5×10^{-2} kg
4 a (i) 0.29 kg   (ii) 0.12 kg   **b** 2.3×10^3 kg m^{-3}

5.2
1 1.2 kPa   **2** 120 kN   **3** 13 kN
4 a (i) 240 kPa   (ii) 3.6 kN   **b** 0.5 mm

5.3
1 a 0.40 m   **b** 12.5 N
2 a 20 N   **b** 100 mm   **c** 200 N m^{-1}
3 a 40 N   **b** 200 mm
4 a 12.3 N m^{-1} **b** 8.8×10^{-2} J   **c** 2.2 N

5.4
1 1.0×10^9 Pa
2 1.3×10^{11} Pa
3 a 9.4×10^8 Pa   **b** 1.2×10^{-2} m

5.5
1 a 3.3×10^6 Pa   **b** 2.8×10^{-4} m   **c** 0.21 J
2 a 2.3 mm   **b** 1.7×10^{-2} J
3 a 470 kN   **b** 47 J
4 a 10.5 J   **b** 3.5 J

Chapter 6

6.1
1 a (i) 3.5 C   (ii) 210 C   **b** (i) 3.0 A   (ii) 0.15 A
2 a 3.8×10^{14} **b** 1.9×10^{21}
3 b 1.1×10^{-25} kg
4 a 800 C   **b** (i) 1600 s   (ii) 8000 s

6.2

1 a 29 kJ   **b** 720 J
2 a 2 A   **b** 22 kJ
3 a (i) 48 kJ   (ii) 3.5 A   **b** 5 A
4 a 12 kJ   **b** 4.5 W   **c** 2700 s

6.3

1 a 6.0 Ω, 10 V, 0.125 mA, 160 Ω, 2.5 mA   **b** 7.5 Ω
2 31 Ω   **3** 0.11 mΩ   **4 a** 1.8×10^{-6} Ω m   **b** 33 mm

6.4

1 a 0.25 A, 12 Ω
2 a 0.03 mA   **b** 0.38 mA
4 a 30.4 Ω   **b** 46 °C

Chapter 7

7.1

1 a 1.0 A, 4.0 A   **b** 5.0 A   **c** 30 W
2 b (i) 2.0 V   (ii) 0.20 A
3 b (i) 4.0 V   (ii) 2.0 V   (iii) 10 Ω
4 a 3.6 V   **b** 30 Ω

7.2

1 a 16 Ω   **b** 3.0 Ω   **c** 4 Ω
2 a 2 Ω   **b** 6 Ω   **c** 1.0 A   **d** 4.0 W
3 a 3.6 Ω
  b 0.83 A
  c 2 Ω: 0.5 W; 4 Ω: 1.0 W; 9 Ω: 1.0 W,
  d 2.5 W
4 a 14.4 W   **b** 2.4 Ω

7.3

1 a 6.0 Ω   **b** 2.0 A   **c** 3.0 V   **d** 9.0 V
2 a 0.5 A   **b** 1.25 V   **c** 0.63 W   **d** 0.13 W
3 a 2.0 Ω   **b** 1.5 V
4 12 V, 2 Ω

7.4

1 a 12.0 Ω   **b** 0.25 A
  c 4 Ω : 0.25 A, 1.0 V; 24 Ω: 0.08 A, 2.0 V;
    12 Ω: 0.17 A, 2.0 V
2 a 20.0 Ω   **b** 1.05 A   **c** 1.05 A, 15.8 V
3 a (i) 2.0 W   (ii) 2.0 W   **b** (i) 2.0 W   (ii) 8.0 W
4 a Q: 0.6 V, 0.06 mA; P: 2.4 V, 0.48 mA
  b P: 0.6 V, 0.12 mA: Q: 2.4 V, 0.24 mA

7.5

1 a 1 kΩ: 0.75 V; 5.0 kΩ: 3.75 V
  b 1 kΩ: 1.3 V; 5.0 kΩ: 3.2 V
3 a (i) 0.5 A   (ii) 8.0 Ω: 4.0 V; 4.0 Ω: 2.0 V
  b (i) 3.0 V   (ii) 4.0 V
4 a (i) 2.8 V   (ii) 6.4 kΩ

Chapter 8

8.1

1 a N   **b** NW
2 a S   **b** sudden switch to N
3 a clockwise   **b** → N → W suddenly
4 a field due to coil along axis
  b 90° if coil field >> Earth's field

8.2

1 a S   **b** W   **c** N
2 a N → S   **b** vertical up

8.3

1 0.14 T
2 2.0×10^{-2} N, East
3 a 2.4×10^{-2} N, East
  b 4.5 A, W → E
  c 0.20 T vertically up
  d South, 8.0×10^{-3} N
4 a 58 μT   **b** 6.5×10^{-5} N, due West

Chapter 9

9.1

1 b 5.1×10^{14} Hz   (ii) 1.5×10^6 Hz
3 a 7.0×10^{14} Hz   **b** 4.6×10^{-19} J
4 a 4.7×10^{14} Hz, 3.1×10^{-19} J   **b** 4.8×10^{15}

9.3

2 a (i) 6.7×10^{14} Hz, 4.4×10^{-19} J
    (ii) 2.0×10^{14} Hz, 1.3×10^{-19} J
3 a 1.7×10^{14} Hz   **b** 2.7×10^{-19} J
4 a 3.1×10^{-19} J   **b** 1.6×10^{-19} J   **c** 2.5×10^{14} Hz

9.4

2 a 1.6×10^{12}
3 a 3.4×10^{-19} J
  b 1.5×10^{15}
  c 2.5×10^{12}
4 a (i) 4.0×10^{14} Hz   (ii) 2.7×10^{-19} J,
  b 2.7×10^{-19} J

9.5

3 a 9.1×10^{-11} m   **b** 2.0×10^{-15} m
4 a 1.3×10^{-27} kg m s^{-1}, 1.5×10^3 m s^{-1}
  b 1.3×10^{-27} kg m s^{-1}, 0.78 m s^{-1}

Chapter 10

10.2

1 a 0.10 m   **b** 1.9×10^{-2} m
2 a 10 GHz   **b** 5.0×10^{14} Hz
3 1.0 V, 1.0 kHz
4 a (i) amplitude = 12 mm, wavelength = 77 mm
    (ii) 180°   (iii) 270°
  b +12 mm

10.5

2 a 2.0 m   **b** (i) 180°   (ii) 225°   (iii) 0
4 b 30 mm

10.6

1 a 1.6 m   **b** 410 m s^{-1}
2 a 0.4 m   **b** 0.53 m

10.7

1 a 2.40 m, 140 Hz   **b** 425 Hz
2 a 2.0 m   **b** 57 Hz
3 a 71 Hz   **b** 142 Hz
4 68–680 Hz

Chapter 11

11.1

3 a (i) 19° (ii) 35°   **b** (i) 59° (ii) 75°
4 a 1.53   **b** 34°

11.2

1 a (i) 25 m   (ii) 33 m   **b** 18°
2 a (i) 1.97×10^8 m s^{-1} (ii) 2.26×10^8 m s^{-1}
3 a 25°   **b** 35°
4 b (ii) 74°

11.3

1 b (i) 41°   (ii) 49°
2 b (i) 34°   (ii) 34°
3 a 65°   **b** (i) 30°   (ii) 45°

11.4

2 550 nm   **3** 0.9 mm   **4** 0.75 m

11.5

3 1.1 mm

Chapter 12

12.1

1 a (i) 0.500 m   (ii) 320 cm   (iii) 95.6 m
b (i) 450 g   (ii) 1.997 kg   (iii) 5.4×10^7 g
c (i) 2.0×10^{-3} m^2 (ii) 5.5×10^{-5} m^2
(iii) 5.0×10^{-6} m^2
2 a (i) 1.5×10^{11} m   (ii) 3.15×10^{-7} s
(iii) 6.3×10^{-7} m   (iv) 2.57×10^{-8} kg
(v) 1.5×10^5 mm   (vi) 1.245×10^{-6} m
b (i) 35 km   (ii) 650 nm   (iii) 3.4×10^3 kg
(iv) 870 MW ($= 0.87$ GW)
3 a (i) 20 m s^{-1} (ii) 20 m s^{-1} (iii) 1.5×10^8 m s^{-1}
(iv) 3.0×10^4 m s^{-1}
b (i) 6.0×10^3 Ω   (ii) 5.0 Ω   (iii) 1.7×10^6 Ω
(iv) 4.9×10^8 Ω   (v) 3.0 Ω
4 a (i) 301   (ii) 2.8×10^9 (iii) 1.9×10^{-23}
(iv) 1.2×10^{-3} (v) 2.0×10^4 (vi) 7.9×10^{-2}
b (i) 1.6×10^{-3} (ii) 5.8×10^{-6}
(iii) 1.7   (iv) 3.1×10^{-2}

12.2

1 a 1.57 m   **b** (i) 1.57 m   (ii) 1.18 m   (iii) 0.39 m
2 a (i) 68°   (ii) 41°   (iii) 22°   (iv) 61°
b (i) 17 cm   (ii) 16 cm   (iii) 5 mm   (iv) 101 cm
3 a 49 mm   **b** (i) 35 km   (ii) 31°
4 a (i) 3.9 N, 4.6 N   (ii) 3.4 N, 9.4 N   (iii) 4.8 N, 5.7 N
b 4.0 N, 30° to 3.5 N

12.3

1 a 0.2   **b** 0.1, 0.25   **c** < 0.1
2 b (i) 2   (ii) -1 (iii) 4.25   (iv) $\frac{1}{3}$
3 b (i) 0.25   (ii) 2.8   (iii) 0.5   (iv) 4
4 a 3.1×10^{-7} m^2
b 6.2×10^{-3} m
c 2.5 s
d 17 m s^{-1}

12.4

1 a (i) 3   (ii) -3 (iii) 1
b (i) -4 (ii) 8   (iii) 2
c (i) -1 (ii) 5   (iii) 5
d (i) -1.5 (ii) 3   (iii) 2
2 a (i) $y = 2x - 8$ (ii) -8
b (i) 3 m s^{-2} (ii) 5 m s^{-1}
3 a $(1, 4)$ **b** $y = 4x$
4 a $x = 2, y = 0$
b $x = 3, y = 5$
c $x = 2, y = 0$

12.5

3 b Gradient $= \dfrac{A}{\rho L}$
4 b (ii) $u = y$-intercept, $\frac{1}{2} a$ = gradient

12.6

1 b (i) gradient   (ii) area under line
2 b Resistance $= \dfrac{1}{\text{gradient}}$
3 b The power is constant and is represented by the gradient of the line.
4 b (i) **1** acceleration **2** distance fallen   (ii) velocity

Index

Bold page references refer to illustrations, figures or tables.

Acknowledgements

I would like to thank my family for their support in the preparation of this book, particularly to my wife, Marie, for secretarial support and cheerful encouragement. I am also grateful to the publishing team at Nelson Thornes, in particular to Beth Hutchins who initiated the project and to Sarah Ryan who coordinated the production of the book. I also thank OCR for their support.

Jim Breithaupt

Photograph acknowledgements

The author and publishers are grateful to the following:

Ann Ronan Picture Library: 11.17
Corel 487 (NT): 1.23; Corel 671 (NT): 4.12, 5.5; Corel 677 (NT) 6.6 (stereo); Corel 771 (NT): 3.17;
Digital Vision 6 (NT): 1.34
Illustrated London News: 9.9;
Last Resort Picture Library: 10.32;
Martyn Chillmaid: 6.7, 8.13;
Photodisc 4 (NT) 6.6 (computer), 10.6; Photodisc 17 (NT);
Science Photolibrary: 3.25, David Parker 1.15, 11.10, Keith Kent 3.14b, TRL Ltd 3.22, 3.23, Peter Menzel 6.4, Rosenfeld Images Ltd 9.4,
Picture research by johnbailey@axonimages.com

Cover Photo by Plasma Globe by Alfred Pasieka/Science Photo Library